Water Policy in Texas

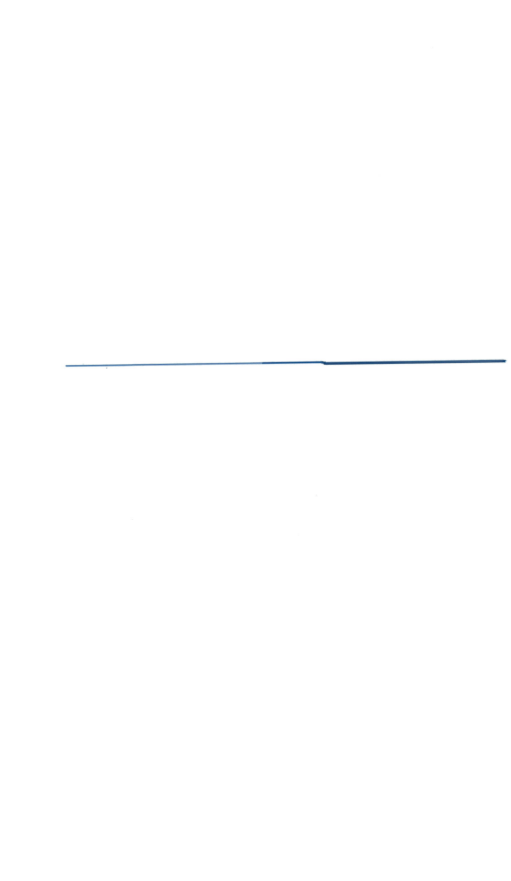

Water Policy in Texas
Responding to the Rise of Scarcity

EDITED BY

Ronald C. Griffin

RFF PRESS
RESOURCES FOR THE FUTURE

Washington, DC • London

First published in 2011 by RFF Press, an imprint of Earthscan

Earthscan LLC, 1616 P Street, NW, Washington, DC 20036, USA
Earthscan Ltd, Dunstan House, 14a St Cross Street, London EC1N 8XA, UK
Earthscan publishes in association with the International Institute for Environment and Development

For more information on RFF Press and Earthscan publications, see www.rffpress.org and www.earthscan.co.uk or write to earthinfo@earthscan.co.uk

ISBN: 978-1-93311-589-4

Copyedited by Joyce Bond
Typeset by OKS Press Services
Cover design by Maggie Powell

Library of Congress Cataloging-in-Publication Data

Water policy in Texas : responding to the rise in scarcity / edited by Ronald C. Griffin.
    p. cm.
  Includes bibliographical references and index.
    ISBN 978-1-933115-89-4 (hardback : alk. paper) 1. Water-supply–Government policy–
Texas. 2. Water-supply–Texas–Management. 3. Water resources development–Texas.
  I. Griffin, Ronald C.
  HD1694.T4W38 2010
  333.91009764–dc22   2010002841

A catalogue record for this book is available from the British Library

At Earthscan we strive to minimize our environmental impacts and carbon footprint through reducing waste, recycling and offsetting our CO_2 emissions, including those created through publication of this book. For more details of our environmental policy, see www.earthscan.co.uk.

Printed and bound in the UK by TJ International, an ISO 14001 accredited company.
The paper used is FSC certified and the inks are vegetable based.

About Resources for the Future *and* RFF Press

Resources for the Future (RFF) improves environmental and natural resource policymaking worldwide through independent social science research of the highest caliber. Founded in 1952, RFF pioneered the application of economics as a tool for developing more effective policy about the use and conservation of natural resources. Its scholars continue to employ social science methods to analyze critical issues concerning pollution control, energy policy, land and water use, hazardous waste, climate change, biodiversity, and the environmental challenges of developing countries.

RFF Press supports the mission of RFF by publishing book-length works that present a broad range of approaches to the study of natural resources and the environment. Its authors and editors include RFF staff, researchers from the larger academic and policy communities, and journalists. Audiences for publications by RFF Press include all of the participants in the policymaking process—scholars, the media, advocacy groups, NGOs, professionals in business and government, and the public. RFF Press is an imprint of **Earthscan**, a global publisher of books and journals about the environment and sustainable development.

Resources for the Future

The RFF Press Water Policy Series

Books in the *RFF Press Water Policy Series* are intended to be accessible to a broad range of scholars, practitioners, policymakers, and general readers. Each book focuses on critical issues in water policy with the mission to draw upon and integrate the best scholarly and professional expertise concerning the physical, ecological, economic, institutional, political, legal, and social dimensions of water use. The interdisciplinary approach of the series, along with an emphasis on real world situations and on problems and challenges that recur globally, are intended to enhance our ability to apply the full body of knowledge that we have about water resources—at local, country, regional, and international levels.

We welcome new contributions to the series. For editorial queries about the *RFF Press Water Policy Series*, please write to *waterpolicy@rff.org*

Contents

Dedicated to Michele Zinn
for her numerous years of superb assistance

Foreword

"Till taught by pain, men really know not what good water's worth."
—Lord Byron, *Don Juan, Canto II, 1819*

B yron's sobering words resound the fundamental truth that water is essential to the sustenance—indeed, the existence—of life. The availability and quality of water are inextricably linked to the health and well-being of all living things. How a society allocates, manages, uses, and protects its water resources invariably determines and defines the quality and vitality of its socioeconomic well-being.

From the days of the first settlers, water has guided Texas's progress and shaped its destiny. The laws, policies, and decisions Texas has made relating to its water resources from the days of basic wilderness to today's twenty-first-century global economic powerhouse have defined the state's past and present, and they will well determine what Texas will be for generations to come.

With its ever common frontierish approaches, in combination with a cultural and political ethic well rooted in private property rights, Texas has been a practical laboratory for testing unique or nonstandard ideas and methods for water resource management. The importance of the state's experiences in water management comes not only from their uniqueness, but also from their severity and diversity. Texas continues to face significant water issues because many of its current water management policies and practices are remnants of an era long past, when its population and water demands were significantly smaller and it was inconceivable that the demand for water would exceed its supply. These policies and practices too often have perpetuated a false sense of security and inhibited any appropriate undertaking of comprehensive, long-term planning and policy development until forced to do so by crisis. Thus, as is in many other parts of the world, institutional changes pertaining to water remain an often unsettled and ever-evolving story in Texas.

All the while, the state has consistently experienced considerable population growth, residing in sprawling urban and suburban areas. It has extensive industrial complexes, vast agricultural regions, diverse water-dependent habitats for fish and wildlife, and a significant water-based recreation sector. Frequently occurring droughtlike conditions and other climate-related variability in water supplies present a strong potential for extreme stress on water resources and magnify all the intense competitions for water.

Texas has one of the world's most robust economies, but if sound, scientifically based water management strategies are not implemented, it could face serious social, economic, and environmental consequences. The state has learned through practical experience that it must arduously protect the quality of its water and judiciously conserve its uses as demands continue to grow. It must ensure that its water policies and management programs are flexible enough to meet changing demands and supplies. They must be established with enough foresight to reasonably predict or anticipate future demands and supplies, and they must be meaningfully cognizant of all the diverse interests that come to bear, or will come to bear, on water supplies and demands. Texas understands that it cannot afford policies and management programs whereby groundwater usage becomes adversarial to surface-water usage, and vice versa; municipal, industrial, and agricultural usages conflict; economic development and growth are put at odds with protection of the environment and conservation of natural resources; or one region of Texas is placed in an antagonistic position toward another.

Yet despite the sometimes uniqueness of its historical and present water management institutions and approaches, Texas nevertheless shares significant commonalities with other water-scarce areas worldwide. Thus, what the state has learned and applied in its 170-year existence, both positively as opportunities and negatively as problems, offers instructive answers and examples to other jurisdictions similarly situated.

As part of the RFF Press Water Policy series, this volume, *Water Policy in Texas: Responding to the Rise of Scarcity*, adds to the growing inventory of the various water management models that are in operation worldwide and offers a meaningful tool to help educate those anywhere who are interested in the further evolution of water policy. It presents a comprehensive account of Texas's unique laws, policies, approaches, and experiences in water management and offers valuable insight into and examples of the numerous and diverse challenges, opportunities, and problems the state has faced, as well as informative lessons learned. But rather than attempt to duplicate available information or comprehensively cover the entire scope of Texas water management, the book instead has selected as its basis the more substantial topics pertaining to water scarcity. Experts, mostly longtime students of the state's water issues and policies, have authored the various chapters on key Texas water topics. Their learned writings span the crucial details of their specific subjects, making the textual materials accessible to a broad policy-oriented and nonacademic audience. As specific water management policies and issues are brought forward by the authors, some different, nontraditional thinking emerges, borne by the unique ways Texas manages its water resources.

Texas water management, current and historical, reflects an evolutionary confluence of climatology, hydrology, geology, history, cultural ethics, socio-economics, and politics, thus leading to a unique style of water management systems and experiences that form a truly instructive model. *Water Policy in Texas:*

Responding to the Rise of Scarcity provides both an invaluable case study into that model and an educational text to develop analytical tools for water management for students, policymakers, and professionals worldwide.

Larry R. Soward
Commissioner, Texas Commission on Environmental Quality
2003–2009

Editor and Contributors

Ronald C. Griffin is professor of water resource economics at Texas A&M University, where he has been a researcher and teacher for 30 years. He is the author of *Water Resource Economics: The Analysis of Scarcity, Policies, and Projects* and is currently an editor of the journal *Water Resources Research*. He specializes in water studies pertaining to demand, pricing, policy, marketing, and cost−benefit analysis.

John B. Ashworth is a senior consultant with the groundwater consulting firm of LBG-Guyton Associates in Austin, Texas, and is a licensed geoscientist in the state of Texas. He currently directs the firm's water management planning activities and is the project manager for two of the state's 16 water-planning regions. Prior to his employment with LBG-Guyton, John spent 23 years with the Texas Water Development Board, where he was involved with the agency's statewide assessment of groundwater availability and in the development of previous state water plans.

Andreas Gondikas is a chemical engineer and master of engineering management by training and is currently a Ph.D. candidate in the Civil and Environmental Engineering Department at Duke University. His research interests focus on the fate and transport of natural and synthetic nanoparticles in the environment. He has experience in environmental consulting as a project engineer, working on wastewater treatment, water, and sewer projects.

David Jassby is a Ph.D. candidate in the Department of Civil and Environmental Engineering at Duke University. He has an M.S. in environmental engineering from the University of California−Davis and a B.Sc. in microbiology from Hebrew University. His research focuses on nanotechnology applications in environmental settings.

Ric Jensen led efforts of the Texas Water Resources Institute to communicate research-based information to the public from 1985 to 2008. He has conducted oral interviews to develop a history of the Pecos River of Texas. Since 2008, he has been an associate professor in the Department of Contemporary Media and Journalism at the University of South Dakota. Jensen's scholarly interests focus on how changes in technology affect communication and include public relations case studies about the environment, sports, politics, and ethics.

Jeffrey W. Johnson is an assistant professor in the Department of Agricultural and Applied Economics at Texas Tech University, where he is the director of farm operations and associate director of CASNR Water Center. His research interests include the economics of water resources, with an emphasis on groundwater policy and irrigation economics.

Ronald Kaiser is chair of the Texas A&M University graduate water degree program and a professor of water law and policy. His research focuses on law, marketing, environmental flows, and groundwater management and conservation. A number of his research recommendations have been incorporated into Texas water law.

Mary E. Kelly is senior counsel for the Center for Rivers and Deltas at Environmental Defense Fund. She has more than 20 years of experience as an environmental lawyer, having worked in private practice and the not-for-profit sector. She manages major Environmental Defense Fund projects to protect and restore habitat, rivers, and coastal deltas across the United States. She has specialized in water law and U.S.-Mexico binational water management during much of her career.

Andrew J. Leidner is a Ph.D. candidate at Texas A&M University, funded through the Rio Grande Basin Initiative and the Texas Water Resources Institute, and a project representative for NRS Consulting Engineers. His research interests include water resource economics and management, with an emphasis on advanced water technologies.

Kathy Alexander Martin holds a Ph.D. in aquatic resources from Texas State University. Her research focuses on the effectiveness of water management institutions in the Rio Grande basin. She has coauthored publications on interbasin transfers and water availability modeling, and she is a founder and an editor of the *Texas Water Journal*.

Paul Montagna is the endowed chair for ecosystem studies and modeling at the Harte Research Institute for Gulf of Mexico Studies at Texas A&M University–Corpus Christi. He is also a professor of environmental science and the coordinator of the coastal and marine system science doctoral program. His research focuses on coastal management, benthic processes, ecoinformatics, ecosystem modeling, environmental flows, and integrating natural science and socioeconomics to improve coastal decisionmaking. He has been a member of the Texas Environmental Flows Science Advisory Committee since 2004.

Ben Vaughan is a visiting assistant professor of economics at Trinity University in San Antonio, Texas. His previous work on Texas water includes a chapter on the Edwards Aquifer in *Water Marketing: The Next Generation* and service on two legislative boards tasked with shaping Texas water policy's environmental impact.

Todd Haydn Votteler is the executive manager of intergovernmental relations and policy for the Guadalupe-Blanco River Authority. U.S. District Court Judge Lucius Bunton appointed him as the federal special master for the Endangered Species Act litigation over the Edwards Aquifer, *Sierra Club v. San Antonio.* He was the federal court monitor's assistant during the Endangered Species Act litigation *Sierra Club v. Babbitt.* Votteler is the executive director emeritus of the Guadalupe-Blanco River Trust and is currently chair-elect of the Texas Land Trust Council. He is a founder and the editor-in-chief of the *Texas Water Journal.*

George Ward is a research scientist at the Center for Research in Water Resources of the University of Texas. He specializes in hydrodynamics and transport processes operating in natural fluid systems. His work includes numerous studies in the coastal and nearshore environments, particularly the analysis and modeling of circulation of the bays and estuaries of Texas. He has been a member of the Science Advisory Committee on Environmental Flows for the State of Texas since 1994.

Mark R. Wiesner is the John L. Meriam Professor of Civil and Environmental Engineering at Duke University, where he also serves as the director of the Center for the Environmental Implications of Nanotechnology. Wiesner's research examines the fate, transport, and impacts of nanomaterials in the environment and has pioneered the application of membrane processes to water treatment. Wiesner is coeditor of *Membrane Processes for Water Treatment and Environmental Nanotechnologies*, and associate editor of the journals *Nanotoxicology* and *Environmental Engineering Science.*

David B. Willis is an associate professor of water resource economics at Clemson University and has been a researcher and teacher for 20 years. His research primarily focuses on western water resource issues, and he specializes in developing interdisciplinary water policy models designed to aid policymakers in achieving an economically efficient allocation of surface water and groundwater among competing uses under supply scarcity.

Yao Xiao is a fourth-year Ph.D. student of environmental engineering at Duke University. He earned a master's degree at Tsinghua University of China, and his thesis was on biological wastewater treatment. His Ph.D. research focuses on the fate and transport of engineered nanoparticles at the Center for the Environmental Implications of NanoTechonology at Duke University.

Acronyms and Abbreviations

ac-ft	acre-foot, acre-feet
AdVE	advanced vapor-compression evaporation
AOP	advanced oxidation processes
AWWA	American Water Works Association
BECC	Border Environmental Cooperation Commission
BOR	U.S. Bureau of Reclamation
Ca^{2+}	calcium ion
cfs	cubic feet per second
CILA	Comision Internacional de Limites y Aguas (Mexican section of the International Boundary and Water Commission)
Cl^-	chloride ion
CO_2	carbon dioxide
CPMP	Critical Period Management Plan
DMI	domestic, municipal, and industrial
DOJ	U.S. Department of Justice
EAA	Edwards Aquifer Authority
EARIP	Edwards Aquifer Recovery Implementation Program
EBPR	enhanced biological phosphorus removal
ED	electrodialysis
EDR	electrodialysis reversal
EFAC	Environmental Flows Advisory Commission
EFAG	Environmental Flows Advisory Group

EPA	U.S. Environmental Protection Agency
ESA	Endangered Species Act
EUWD	Edwards Underground Water District
FO	forward osmosis
FWS	U.S. Fish and Wildlife Service
GAC	granular activated carbon
GBFIG	Galveston Bay Freshwater Inflows Group
GBRA	Guadalupe-Blanco River Authority
GCD	groundwater conservation district
GMA	groundwater management area
gpcd	gallons per capita per day
GPD	gallons per day
H_2O_2	hydrogen peroxide
HCP	Habitat Conservation Plan
HO·	hydroxyl radicals
IBC	International Boundary Commission
IBWC	International Boundary and Water Commission
ITP	incidental take permit
IWT	Innovative Water Technologies group of the TWDB
km^2	square kilometer(s)
kWh	kilowatt-hour(s)
L	liter(s)
LCRA	Lower Colorado River Authority
LEPA	low energy precision application
LGWSP	Lower Guadalupe Water Supply Project
m^2	square meter(s)
m^3	cubic meter(s)
m^3s^{-1}	cubic meters per second
MBR	membrane bioreactor
MD	membrane distillation
MEE	multiple effect evaporation
MF	microfiltration

Mg^{2+}	magnesium ion
MGD	million gallons per day
MSF	multistage flash
MSF-M	multistage flash: mixing system
MSF-OT	multistage flash: once-through system
MUD	municipal utility district
MVC	mechanical vapor compression
Na^+	sodium ion
NADBank	North American Development Bank
NAFTA	North American Free Trade Agreement
NAS	National Academy of Sciences
NF	nanofiltration
NH_3	ammonia
NH_4HCO_3	ammonium bicarbonate
NPV	Net present value
NRC	National Research Council
O_3	ozone
PAC	powdered activated carbon
PEM	polymer electron membrane
ppt	parts per thousand
RIP	Recovery Implementation Program
RO	reverse osmosis
SAC	scientific advisory committee
SARA	San Antonio River Authority
SAWS	San Antonio Water System
SB	Senate Bill
SCTWAC	South Central Texas Water Advisory Committee
SEMARNAT	Mexican Secretaria de Medio Ambiente y Recursos Naturales
SMRF	San Marcos River Foundation
sq km	square kilometer(s)
TAMU	Texas A&M University
TCEQ	Texas Commission on Environmental Quality

TGPC	Texas Groundwater Protection Committee
THP	Texas High Plains
TMDL	total maximum daily load
TNRCC	Texas Natural Resource Conservation Commission
TPWD	Texas Parks and Wildlife Department
TRRC	Texas Railroad Commission
TRWD	Tarrant Regional Water District
TSSWCB	Texas State Soil and Water Conservation Board
TWC	Texas Water Commission
TWDB	Texas Water Development Board
UF	ultrafiltration
USACE	U.S. Army Corps of Engineers
USGS	U.S. Geological Survey
USIBC	U.S. section of the International Boundary Commission
USIBWC	U.S. section of the International Boundary and Water Commission
UV	ultraviolet radiation
UWCD	underground water conservation district
WAM	Water Availability Model
WCID	water control and improvement district
yr	year

CHAPTER 1

Experiments in Water Policy

Ronald C. Griffin

A major principle of American water management is that each of the 50 states is allowed to choose its own path, especially its guiding doctrines for how water is to be allocated.[1] So states establish unique water policies. Each forms its individual body of law for administering both groundwater and surface water, often borrowing ideas from earlier-acting states. Because the U.S. government, through its deferral to state authority on most water use matters, has passed water ownership (at least much of it) to the states, any state can subsequently pass along this authority to local agencies or individual water users. States have many options.

With all of their policy selections, states can and do favor particular water uses over others. Sometimes this favoritism is explicit, such as when states establish preference orderings that lay out sectoral rankings concerning water use priorities.[2] Sometimes the favoritism is rooted in traditions that broadly permeate rules and attitudes. For example, Texas has tended to establish rules favoring diversionary uses (out-of-water body) over instream, in-lake, and in-aquifer uses such as habitat, recreation, and pumping lift protection.[3]

Arguably, no state has taken fuller advantage of U.S. legal latitudes than has Texas. As compared with other states and countries, Texas has been a laboratory for testing nonstandard ideas for how water might be managed. Moreover, these policy decisions have been strongly influenced by a cultural ethic that is respectful of private property rights. The implications of these combined forces are fascinating. No other state espouses a baseline legal code in which landowners can withdraw as much groundwater as they like. No other state has embraced the buying and selling of water to the high degree Texas has. Among the 17 western states, where both the U.S. Army Corps of Engineers and the U.S. Bureau of Reclamation were major players in twentieth-century reservoir development, no other state has a federal water presence that is as diminished as in Texas. That is, though these agencies erected major reservoirs and other water control facilities in the state, legal control over the created water rights lie within the state. Furthermore, with considerable authority wielded by individual utilities, districts, and individual right holders, the water management activities enacted in Texas arise from a more decentralized process than is normally witnessed elsewhere.

These and other distinctions raise a serious question made even more interesting by the physical commonalities Texas shares with other water-scarce areas across the globe: What has been learned in the Texas lab—either positively or negatively—that might be usefully applied elsewhere? One purpose of this book is to uncover some answers to this question. Because of the unique institutions that are sometimes applied in Texas, revealing examples of policy opportunities and problems have emerged.

Another purpose of this book is to assemble key status elements for the current state of water management in Texas. Where are we with respect to the larger water issues? How are things being done? What is changing? What are the options and what outcomes are possible for Texas? Instead of attempting to duplicate available information or comprehensively cover the entire Texas waterscape, the more substantial topics pertaining to water scarcity have been selected.

Western states such as Texas commonly produce a state-level water plan every several years, with the general goals of inventorying problems, proposing solution packages, and coalescing political support for undertaking action. Texas's most recent plan (2007) is available online, and it is a storehouse of information that need not be replicated here.[4] The Texas water plan departs from older planning approaches in the method of its creation, but its recommendations lie on the same evolutionary ladder as its predecessors. Policy emphasis remains on the construction of proposed reservoirs and other structural measures; yet actual construction has been slow relative to calls for it by water plans. Billions of dollars are said to be needed for infrastructural advance.[5] To a large extent, the reservoir development focus of past state water plans is well aligned with prevailing public sentiment. The majority opinion of citizens and leaders is that we must always be developing new water supplies to accommodate growth. In Texas, further growth is thought to be important for increased well-being.

To reach beyond commonly held beliefs, experts who are mostly longtime students of Texas water issues have been invited to produce original writings on key Texas water topics for this book. Most of the authors have participated in the creation of new information through research on their subjects. Others have been engaged in actual decisionmaking processes during which notable policies have been advanced. Some have done both. They have been asked to write foundation pieces that span the crucial details of their subjects, so that this material can be accessible to a broad audience. They have also made a concerted effort to connect readers with additional information sources via thorough referencing, including URLs where possible. Not surprisingly, as specific water management issues are brought forward in this book, some outside-the-box, different thinking will emerge, spurred by the unique ways that water is sometimes managed in Texas.

The remainder of this introduction provides an overview of the more unique features of Texas as a water-using region and what this book offers in its forthcoming chapters. But first it is necessary to clarify a few terms in order to move beyond the simplistic rhetoric that often clouds discussions about water. This book confronts hard issues, making a uniform terminology more important.

A HEADS-UP ON SEMANTICS AND JARGON

When people have different visions of the same term, dialogue becomes empty or frustrated, and communication is hampered. We are attempting to avoid such pitfalls in this book, especially for the following terms.

Everyone seems to agree that "sustainability" in water use is a desirable goal, but available definitions of the word are inconsistent, unmeasured, and too soft to guide decisionmaking, so we avoid the term in this book whenever possible. "Sustainable water use" is a commonly heard goal in all Texas forums, even as political and administrative decisionmaking continues to proceed along unsustainable paths. Perhaps the word itself enables these contradictions through its ambiguity and invitation for loose use. To encourage more meaningful dialogue, we abandon the word. This does not imply that the underlying ethic of sustainability is opposed here.

In both public discussion and written media and reports, water "needs" are commonly claimed when the truth almost always lies closer to water "wants." To exercise greater caution in the forthcoming chapters, "needs" is reserved for primarily biological requirements, as in the cases of drinking water and water for species-supporting habitat. Only in such cases is the term "need" arguably accurate.

In a similar vein, water "demand" is not applied here as a substitute for words such as "use" or "consumption." Whereas water use or consumption is adequately described by a quantity, such as a certain number of acre-feet or gallons, demand is a broader term that encompasses the idea of variable usage as conditions change. For example, demand incorporates household or business behavior whereby people modify their water usage as water scarcity changes. Water demand incorporates the functional dependence of water use on its determining factors, such as weather or water rates. Therefore, demand is harder to know than is use or consumption, and use or consumption does not reveal everything we would like to understand about demand.

KEY CONDITIONS

The importance of Texas's experiences in water management derives from not only their uniqueness, but also their severity and diversity. As is true in other parts of the world, institutional change pertaining to water remains an unsettled and evolving story here. During each legislative session of recent years, Texas's governing body has been reshaping water policy with considerable seriousness. The search is clearly on for still better ways of doing things. The stakes are high, as the following few facts begin to suggest:

• Five of the nation's 20 largest cities are in Texas (3 of the largest 10; 6 of the largest 25). Texas is now the second most populous state (behind California) and continues to urbanize at a quick pace.

- Texas has the second-largest land area among American states (behind Alaska) and is larger than many countries (for example, as points of reference, Afghanistan, Ecuador, France, Somalia, and Ukraine are all smaller). Less than 2% of Texas land is federally owned, which is in considerable contrast to more western states.
- Even on a world scale, the Texas economy is significant. Only 15 countries have larger gross national products than Texas's gross state product.
- As is true for other regions where water resource issues are especially problematic, irrigation is the largest use of water in Texas. Among the U.S. states, only California and Nebraska have more irrigated land, and only California's agricultural products have a higher total market value. As in many countries, transitioning from irrigation's dominance is a clear challenge in light of Texas's continuing population and economic growth.
- Texas constitutes almost two-thirds of the relatively arid U.S. border with Mexico. The 1,241-mile Texas border with Mexico is defined by a river, or at least a riverbed (as it may be devoid of flow in particular segments). Not only does a 1944 U.S. treaty with Mexico exist for dividing and managing these waters, but Texas also has official surface-water "compacts" with all of its neighboring states. These are interstate agreements that apportion water, and they have the strength of federal guarantee in that they are enforced by the U.S. judicial system. Yet both treaty and compact compliance has been contentious for Texas, especially as the state is commonly a downstream partner in these arrangements, receiving only what water upper basin jurisdictions do not take. Consequently, considerable understanding has been produced regarding shared water resources, and important treaty experience exists.
- Texas has a long coastline where many rivers and lesser watercourses empty into the Gulf of Mexico. Among the 48 contiguous states, only Florida, California, and Louisiana have longer coastlines. An important environmental concern and, many argue, policy failure is the reduction of freshwater inflows to Texas bays and estuaries consequent to upstream human uses of water. Similar issues arise with respect to instream flow reservations, another topic where Texas policy has been arguably deficient but is evolving. Indeed, it has been said that recent Texas reforms establish an important model.[6]
- With the exception of a single river basin (Rio Grande), Texas does not receive appreciable water supply assistance from snowpack—one of nature's most significant reservoirs. This fact constitutes a scarcity-elevating disadvantage for Texas relative to many other states and countries. Yet in a warming-climate context, this is also a point of possible convergence, as snowpack diminishes in reliability and quantity for other regions.
- Federal agencies have spent significant moneys constructing water reservoirs in Texas over the past 100 years, and these agencies often continue as operators of these large and economically important surface-water impoundments. But unlike with similarly conducted water development in other western states, federal agencies did not gain title to the water rights that were created by these storage facilities. This fact lessens the federal role and lowers the number of political players that must be appeased during policy reform and water

reallocation procedures. It also affects the political activities of groups trying to preserve their favored status as beneficiaries of federal largesse in water control. As compared with more tedious policy deliberation processes in states such as California, the differences can be stark.

CHAPTER-BY-CHAPTER TOPICS

To establish basic foundations pertaining to the water resource conditions of Texas, Chapter 2, by John Ashworth and Ric Jensen, surveys the physical and human dimensions of the state's water issues, observing major challenges that have arisen. Among the noted problems, all of which are growing, are excess demand relative to supply, flood and drought cycles, springflow decline and elimination, groundwater depletion, advancing population, environmental water shortages, and water quality degradation. The state's diversity is quite evident in this survey. A brief history of planning efforts during the past half century also provides context for understanding the development of recent strategies, which were created within a redesigned planning process.

In Chapter 3, Ron Kaiser describes the water laws of Texas and also identifies the roles of the multitude of agencies operating in the state. The chapter's focus is on Texas's version of the prior appropriation doctrine for surface water and its application of the rule of capture for groundwater. A noteworthy historical point is the once confused system of surface-water rights, in which various inconsistent doctrines were impossibly applied at the same time. Major approaches for accomplishing surface-water and groundwater transfers are introduced in this chapter, as are some of the attendant issues. An important topic is recent reform that grafts groundwater districts onto the rule of capture in the state's efforts to repair this depletive doctrine. For readers unfamiliar with this material, it is important background for nearly all of the following chapters.

In Chapter 4, I examine the conduct of Texas water marketing and ratemaking for its policy lessons. Solutions like these may become critical as water-scarce regions confront the impracticality of harnessing more water in environments where the resource is inevitably constrained. The chapter explains the conduct and results of five distinct Texas water markets, and then critiques these practices for their achievements. Parallel attention is devoted to Texas procedures for setting the water rates paid by the clients of water supply organizations. Contemporary market prices and retail rates are indicated in the discussion. Because consumers do conserve more appropriately when rates more completely reflect resource value, socially useful tools are available here, although their subtleties and long-run importance are commonly underappreciated by regulation-minded policymakers.

The purpose of Chapter 5, by Todd Votteler, is to tackle the compelling case of an important water resource: the Edwards Aquifer. Endangered species concerns arising with Edwards-sourced springflow forced Texas to abandon the rule of capture, but only for this aquifer. The legislature replaced this rule with a transferable water rights system in 1993, but it took more than a decade to implement this change, largely because of the long process of documenting water claims and assigning quantified rights. This tale of reform is chronicled here, with additional attention to recent political struggles as dissatisfied parties attempt to soften the new system. The entire transformation has historical consequences, not only for the region, but potentially for other jurisdictions as well, should they find a role model here.

In Chapter 6, Paul Montagna, Ben Vaughan, and George Ward overview the available science background required to appreciate the problems of reserving freshwater inflow for estuary systems. Individual water users see water in a river and wonder why they should not be allowed to use it. During much of Texas's 170-year existence, it has responded to such demands by granting ever-rising rights to take water. Evidence now suggests that it overstepped in allocating water away from estuarial environments. Recently the state instituted measures that begin to correct these deficiencies. At the heart of these remedies is a charge to scientifically assess and quantify the amount of water that should be reserved. This chapter illuminates the task of assessing estuary wellness and its functional association with freshwater inflows. It also provides important details about the complexities of these environments and the manners in which freshwater inflows influence the services estuaries provide.

Mary Kelly visits the inland issue of preserving streamflows for environmental purposes in Chapter 7. Texas avoided meaningful attention to this problem until 1985, as observed here. Progress continued to be slow but was rejuvenated by novel water right applications in recent years. Although maneuvers by environmental interests were rebuffed administratively, they generated additional impetus for policy evolution. Among the reviewed results are directives to assemble experts and commission studies to be used as a basis for crafting water set-asides. Texas can now claim important strides. Water marketing is even part of the solution package. Whether these reforms will produce a substantially improved environment remains unproven, but they demonstrate that low regard for environmental water does not have to be a permanent condition.

Chapter 8, by Kathy Alexander Martin, is dedicated to Texas's formal water relations with its neighboring nation and states. The stories of five interstate water compacts are told. These contractlike divisions of surface water involve all border basins and all border states plus Colorado. Martin describes histories, details, and issues for each compact. Most of the chapter is dedicated to the historical development and consequences of the 1944 treaty with Mexico. This particular treaty speaks to the division of the Rio Grande as well as the not-in-Texas Colorado River of the U.S. Southwest. For each nation, shares are specified for each contributing tributary of the Rio Grande, so these arrangements are seemingly precise. Yet the language and application of the treaty have shortcomings, giving rise to a major dispute in recent years. This too is a significant story.

In Chapter 9, David Willis and Jeff Johnson consider the case of the southern section of the Ogallala Aquifer, which underlies not just northwest Texas, but other Great Plains states as well. Recharge to the Texas portion of this large aquifer is slight, yet this region developed an irrigation-fueled economy founded on groundwater pumping. Depletion is a serious problem, and the limited surface-water alternatives are fully committed. Willis and Johnson perform some exemplary analysis of this issue after first giving an overview of the physical setting and institutional and irrigation history of the region. The authors model two policies under consideration for their likely impacts on the problem, thereby providing an eye-opening look at the consequences of politically acceptable yet light-handed policy measures. Implications for the region and ones like it are dramatic.

Chapter 10, by David Jassby, Andrew Leidner, Yao Xiao, Andreas Gondikas, and Mark Wiesner, surveys the various technology-inspired options for enhancing Texas's water supply in "frontierish" ways. The importance of this topic arises from the confidence many people express in the potential of technology to solve our water scarcity problems. Is this realistic? Approaches discussed here include desalination, wastewater reuse, water harvesting, and water fabrication. The authors identify Texas's progress with some of these approaches and review cost-side evidence and energy use intensiveness.

The book ends, in Chapter 11, with a summary of lessons learned from Texas's experiences and a look toward the future. This synthesis provides an overview of the state's progress in water management and identifies key issues lying on the current frontier.

NOTES

1. There are, however, noteworthy areas in which federal power over water is exerted. These include matters of navigation, Native American rights, endangered species support, federal land operations (e.g., national parks), and hydropower. Where these water applications are important, there are interfaces between state and federal powers to be respected. Yet federal powers are mild in Texas because of the relative (to other western states) absence of these federal activities. Less than 2% of Texas land is federally owned; most Native Americans were displaced beyond the state's borders early in its history; commercial navigation is slight except in the vicinity of Gulf harbors and along the Intercoastal Waterway lying adjacent to the Texas coast; and hydropower production is in the hands of Texas entities (largely river authorities).

2. It must be noted that such priority lists do not find much application in actual decisionmaking, so they are more theoretical than real. Other tools, such as water markets, have been more effective in resolving conflicts. The Texas list of water use preferences, beginning with the highest rank, consists of domestic and municipal, agriculture and industry, mining, hydropower, navigation, recreation and pleasure, and "other beneficial uses" (Texas Water Code § 11.024, http://www.statutes.legis.state.tx.us/Docs/WA/htm/WA.11.htm#11.024, accessed June 2, 2010). What qualifies as a beneficial use is essentially unlimited in the state (Texas Water Code § 11.023).

3. Little consideration of biological water use within natural environments was present in the Texas Water Code prior to the twenty-first-century revisions.

4. Texas Water Development Board, 2007 State Water Plan, http://www.twdb.state.tx.us/wrpi/swp/swp.htm (accessed June 2, 2010).

5. Ibid., ch. 1 Highlights, 2.

6. Lawrence J. MacDonnell, Return to the River: Environmental Flow Policy in the United States and Canada, *Journal of the American Water Resources Association* 45 (October 2009): 1087–99.

CHAPTER 2

Texas Water Resources

John B. Ashworth and Ric Jensen

Texas can be characterized as a land of diversity. Its sheer size (268,000 square miles) encompasses a multitude of geographic and topographic features ranging from coastline to mountainous terrain, marshland to desert habitat, and uninhabited to densely populated landscapes. In some respects, Texas resembles a nation. It has a unique history relative to other U.S. states, and this has generated some institutional differences. For a few years in the 1800s, before joining the United States, Texas *was* a nation. Texas is slightly larger than Ecuador and more than twice the size of Italy. Today's population of 21 million exceeds that of many other countries. From an economic perspective, the gross state product of Texas totaled more than $989 billion in 2005, which, if Texas were an independent nation, would rank its economy sixteenth in the world.[1]

In terms of water use, Texas is like many other states in the U.S. West. Since Anglo-American settlement in the mid-1800s, agricultural irrigation has consistently been the largest use of water. As the population of Texas continues to increase, municipal and industrial water use may surpass the amount of water used by agriculture. Making a smooth transition from rural to urban water use has proven to be challenging from economic, political, and legal points of view.

Water supply sources used to meet demands are also slowly evolving. Early settlers migrating to Texas first established communities along trade routes that depended on reliable water supply sources. At the time of Texas independence from Mexico in 1836, shallow dug wells and springs served as sources of drinking water throughout much of the state. As well drilling and pump technology developed in the early to mid-1900s, more prolific groundwater sources (aquifers) were discovered that became a prevalent source of supply, especially for irrigation use in the dryer western part of the state. Pumping in excess of recharge has caused groundwater tables to fall in many areas, and thus many of the original shallow wells are now dry, and numerous springs have significantly diminished or ceased to flow altogether. Still, deeper wells continue to furnish fresh water to many communities.

Also in the early 1800s, many settlements were initially sited along rivers, only to be abandoned because of the unpredictable nature of floods that swept through their communities. Since that time, dams have been built in many of these flood-prone regions, and the resulting reservoirs have become an important water supply. Today Texas heavily relies on both surface-water and groundwater sources to supply the state's growing water demands.

Texas is actively addressing current and anticipated future water issues through a planning process that starts at the grassroots level. Legislatively mandated procedures are in place to develop strategies that are both physically and economically feasible to meet projected water use levels. However, changes to long-standing water management practices will not come without a challenge in both the legislature and the court system.

Just as water supply development and use have evolved over time, so have the state's water laws transitioned from foreign influences to today's prior appropriation and private ownership institutions. Chapter 3 provides an in-depth look at these water laws, their historical origins, and their governing agencies.

This chapter provides a basic overview of some of the physical characteristics and key issues that influence the availability of water throughout Texas. It discusses current and projected water use trends and the available water resources to satisfy these trends, presents a brief description of the history of water planning and water management in Texas, and gives a perspective on the current process the state employs in water supply planning. The chapter liberally references the Texas Water Development Board's *Water for Texas 2007*, which documents the state's most recent water plan.

PHYSICAL DIVERSITY OF TEXAS

From the salt marshes and barrier islands of the Gulf Coast to the mountainous western region, the geography of Texas spans a wide range of natural features (Figure 2.1).[2] The Gulf Coastal Plains parallel the Gulf of Mexico and include the Coastal Prairies, the Interior Coastal Plain, and the Blackland Prairies. The Edwards Plateau and Grand Prairie form the limestone plateau country stretching from the Dallas–Fort Worth area in north Texas southward to San Antonio and westward to the Trans-Pecos region of the state. A rolling basin floor studded with rounded granite hills characterizes the Central Texas Uplift, and red soils developed from outcropping Permian-age rocks distinguish the North-Central Plains. The High Plains province covering the Texas Panhandle is the southern terminus of the midcontinent Great Plains and is the most extensively irrigated region of the state. The mountainous region of far west Texas is a southern extension of the Basin and Range province and includes the highest elevation in the state, Guadalupe Peak, at 8,749 feet.

Except for the northern Panhandle region, water features form the boundary of much of the state, thus giving Texas its distinctive shape. The Red and Sabine

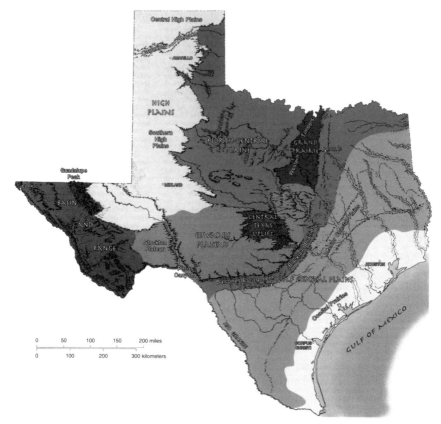

Figure 2.1 *Physiographic map of Texas*

Source: Bureau of Economic Geology, *Physiographic Map of Texas*, University of Texas at Austin, 1996.

Rivers bound the northeast, forming the borders with Oklahoma and Louisiana, respectively; the Gulf of Mexico outlines the southeastern coastline. The Rio Grande forms part of the western boundary with New Mexico and the entire southwestern international boundary with Mexico. Within the state, the land surface generally slopes from higher elevations in the Texas Panhandle and west Texas to lower elevations along the Gulf Coast. Thus, all surface water in Texas that is not evaporated, recharged underground, or withdrawn for local use drains to the Gulf of Mexico.

Ranging from arid in the west to subtropical along the southern Gulf Coast, Texas experiences a wide range in climatic conditions.[3] Moisture-laden tropical breezes traversing inland from the Gulf of Mexico provide the energy that produce spring and fall thunderstorms throughout the state. In addition, hurricanes and tropical storms originating in the Atlantic Ocean often funnel through the Gulf, causing wind and flooding problems from the coastline to deep within the interior. Although the Rocky Mountains, west of the state, tend to limit the severity of Pacific and Arctic winter storms, northerly weather fronts still find their way into the state during the winter and early spring seasons. During summer months, westerly Pacific winds warmed by the high desert plains of Mexico create dry heat waves across much of the state.

The average annual daily high temperature ranges from 70°F (21°C) in the northern Panhandle to 82°F (28°C) in the Lower Rio Grande Valley of south Texas. Relative humidity is highest along the Gulf Coast, at 60%, and progressively decreases inland to less than 35% in the far western region. Precipitation generally decreases as one moves inland from the Gulf Coast, with average annual precipitation ranging from a high of 55 inches at Beaumont in the eastern coastal part of the state to less than 10 inches in the extreme westerly region (Figure 2.2).[4] Throughout much of Texas, especially in the far west, the annual potential for evaporation generally exceeds precipitation. In general, daily variations in temperature and precipitation increase as one moves inland from the Gulf of Mexico. Similarly, the difference between average summer and winter monthly temperatures increases in areas farther away from the Gulf of Mexico.

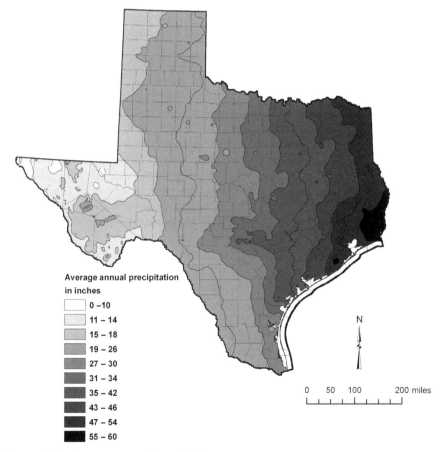

Figure 2.2 *Average annual precipitation*

Source: T. Larkin and G. Bomar, *The Climatic Atlas of Texas*, Texas Water Development Board, 1983.

Surface runoff produced from above-average precipitation events results in a variety of high-flow conditions throughout Texas. Flash floods are created when intense bursts of rain occur within a narrowly bounded drainage basin, resulting in rapidly rising river stages and high-energy currents, and are most common in west Texas and along the southern and eastern Edwards Plateau escarpment. More widespread flooding involves prolonged rainfall events spread out over larger areas, overwhelming the in-channel ability of rivers to capture all the runoff. Rivers may remain at or near flood stage for several days as excess water slowly drains from adjacent lands. The effects of river floods are most often felt on the Coastal Prairies, where land surfaces flatten and river currents slow. Generally speaking, the volume of runoff matches rainfall trends. Average annual runoff is highest in east Texas (more than 18 inches along the coast near Galveston) and lowest in far west Texas near El Paso and in the western portions of the High Plains (only 0.1 inch).

At the drier extreme of the climate spectrum, Texas periodically experiences droughts and water shortages. Over the past decade, increased population and water use in some areas of the state have exacerbated the impact of naturally occurring drought periods. Thus, today water planners often consider drought conditions as periods when it becomes difficult for a region to supply historical amounts of water to key user groups.

At least one period of severe drought has marked each decade since the 1820s.[5] Since weather data started being compiled in the early 1800s, the Trans-Pecos, Lower Rio Grande Valley, and Edwards Plateau have suffered the most droughts. The 1930s drought, associated with the infamous "dust bowl" throughout much of the American Midwest, mainly affected the Texas High Plains and North Plains. Texas's most prolonged drought stretched throughout most of the 1950s and affected the entire state to varying degrees.[6] More recently, severe droughts occurred in 1996 and 1998, resulting in widespread crop failure, significant environmental stress, municipal water shortages, and a number of heat-related deaths.[7] The summer of 2009 produced a dry heat wave that placed much of central and south Texas in an extreme drought status, the most severe on the Palmer Drought Severity Index.

The frequency of droughts and their economic impact is a reason why, in 1997, the state legislature implemented Senate Bill 1, which developed a statewide water supply planning process requiring, among other things, the drafting of drought contingency plans by wholesale and retail public water suppliers and irrigation districts.[8]

How costly are droughts? Billingsley suggests that the 1950s Texas drought of record caused more than $3.3 billion in damages.[9] Similarly, Wilhite described how the widespread impacts of the 1950s drought on Texas forced the federal government to provide assistance for such sweeping issues as water shortages, declining agricultural production, and associated losses of income for rural communities.[10] More recently, agricultural losses were estimated to be about $5 billion from the 1996 drought and $6 billion from the 1998 drought.[11] The Texas Water Development Board (TWDB) estimates that direct costs of water shortages during droughts could total $9 billion annually in 2010 and may escalate to nearly $100 billion per year in 2060.[12]

POPULATION AND WATER USE TRENDS

The population of Texas has grown from 8 million in 1950 to 21 million in 2000, and by 2060, it is projected that nearly 46 million people will reside in the state.[13] The population of 43 of the state's 254 counties and nearly 300 cities, including most major metropolitan areas, is expected to double by 2060. However, not all areas of Texas are anticipated to experience such growth: 45 counties and 137 cities, mostly in west Texas and other rural areas, are projected to remain unchanged or decline in population.

Although significant population increases are expected in Texas, overall water use is predicted to increase by only 27%, to more than 21.6 million acre-feet, as a result of anticipated declines in agricultural irrigation and the more widespread implementation of water conservation programs.[14] Provisions have not been made for this increase, however, especially in light of depleting groundwater supplies, so it remains uncertain whether the projected water use increase will actually materialize.

Although total water use will increase, the water use trends in various segments will experience varying changes in the next 50 years (Table 2.1). Water uses that are more directly related to the urban population of the state (i.e. municipal, manufacturing, and power generation) show the greatest projected increases. Generally across the state, population increases in cities and suburbs are expected to double municipal water use from 4 million acre-feet in 2010 to 8.3 million acre-feet in 2060. On the other hand, irrigated agriculture water use is anticipated to decrease from 10.2 million acre-feet to 8.6 million acre-feet annually.[15]

Although total projected water use is expected to increase in the future, the TWDB anticipates that current supplies will decrease by 18% as existing reservoirs accumulate sediment and aquifers are depleted.[16] The TWDB estimates that unless new supplies are developed or improved management strategies are implemented, Texas could face significant water shortages amounting to 8.8 million acre-feet by 2060, and that more than 85% of the state's population could face water shortages during drought conditions.[17] According to the 2007 state water plan, approximately $30.7 billion in capital cost will be needed by water supply entities and the private sector to meet the state's projected water use levels.[18]

ENVIRONMENTAL WATER NEEDS

Texas includes habitat for nearly 250 species of freshwater fish as well as a diverse assembly of other aquatic animals and plants.[19] Although Texans recognize the tremendous value provided by healthy aquatic habitats, human activities have nevertheless altered the ecology of rivers, streams, and springs.[20] The connection between increased human use of water and the quality of the environment can be seen clearly in many ways. For example, as dams are built to store water for human consumption, existing riparian areas are flooded and downstream riverine areas may be significantly altered. As more water is consumed upstream for human uses, the volume of water that flows downstream into streams and rivers decreases and water quality may suffer. The Texas Parks and Wildlife Department indicates that

Table 2.1 *Water use projections for 2000–2060*

Category	2000	2010	2020	2030	2040	2050	2060
Municipal	4,047,661	4,770,501	5,483,790	6,120,377	6,739,592	7,450,792	8,258,942
Manufacturing	1,559,912	1,825,686	2,004,666	2,163,421	2,319,913	2,452,107	2,578,582
Mining	278,624	270,845	280,815	285,964	276,054	276,931	285,573
Steam-electric	561,394	755,170	886,580	1,030,212	1,174,170	1,339,733	1,533,556
Livestock	300,441	344,495	374,724	381,241	388,243	395,945	404,397
Irrigation	10,228,528	10,345,131	9,980,301	9,585,833	9,206,620	8,843,094	8,556,224
Texas	16,976,560	18,311,828	19,010,876	19,567,048	20,104,592	20,758,602	21,617,274

Source: TWDB, *Water for Texas 2007* (Austin, TX: TWDB, 2007), 122.

fish kills since 1958 have often resulted from a combination of human-induced and natural conditions, including low levels of dissolved oxygen, fast-moving cold fronts, and blooms of toxic algae.[21]

Seven major bay and estuary systems lie along the Texas coast, marking areas where fresh water from rivers mixes with saline water from the Gulf of Mexico. Bays and estuaries are important in that they provide critical habitat for fish, shrimp, oysters, and crabs. In economic terms, bay and estuary systems generate $2.5 billion annually, when the value of fisheries harvests and coastal recreation are considered.[22]

One sign of the loss of aquatic habitat in Texas is the decline in the number of springs that exist statewide. Brune attributes the loss of springs throughout the state and the diminished flow in many more to the lowering of groundwater levels caused by increased pumping.[23] Although many of the larger springs have been negatively affected, two major springs, Comal and San Marcos, still flow from the Edwards Aquifer in central Texas. Each of these springs flows at a rate of more than 15,000 gallons per minute (1,000 liters per second).

The TWDB *Water for Texas 2002* plan recommended that Texas ensure adequate freshwater flows to support aquatic ecosystems in rivers, lakes, bays, and estuaries; however, there was still a lack of consensus about the volume and timing of water releases needed to support these ecosystems.[24] To help address concerns about the effects of water use on the environment, the state legislature established the Texas Environmental Flows Advisory Committee. The committee recommended that a science-based process be developed to examine instream flow needs in major river basins, and that the private sector should be given the opportunity to contribute water rights to the Texas Water Trust to provide more water for environmental needs. Chapters 6 and 7 focus on these topics.

WATER SUPPLY SOURCES

Texas is blessed with a significant supply of water derived from a variety of surface-water and groundwater resources. However, some of these resources are unpredictable in their availability. The hydrologic characterization and development of appropriate management practices for each of these resources are a critical component of planning.

Surface Water

Texas has more than 191,000 miles of streams and rivers, 15 major river basins, and 8 coastal basins (Figure 2.3). On a statewide basis, surface water accounted for roughly 40% of the water used throughout Texas in 2003.[25] In the future, as groundwater resources are exhausted and pumping costs escalate, some regions may develop a greater reliance on surface water.

Widespread flooding along many river segments in the early twentieth century provided an impetus for dam building as a way to manage and store high flows.[26] Texas has 196 major reservoirs (175 with a water supply function), each of which can store more than 5,000 acre-feet of water. Most of the state's 150 dams were

Figure 2.3 *Major river and coastal basins*

Source: Texas Water Development Board, *Water for Texas 2007*, 139.

constructed between 1950 and 1980. The largest reservoir is Toledo Bend, along the Texas-Louisiana border; it can store up to 4.5 million acre-feet.[27]

An estimated 20 million acre-feet of surface water is currently permitted through permanent consumptive surface-water permits. However, many of these rights are not physically available during low-flow conditions. By 2010, only 13.3 million acre-feet per year of surface water are estimated to exist during drought conditions, and of this amount, only 9.0 million acre-feet per year is expected to be physically and legally available for use.[28] Because sediment buildup occurs in many lakes, the volume of existing surface-water supplies is projected to decrease to 8.4 million acre-feet by the year 2060.[29] Sedimentation problems are generally less of a concern in watersheds where small dams have been constructed to trap and hold eroded soils. More than 2,000 small dams have been built in agricultural watersheds throughout Texas.

The Texas Commission on Environmental Quality (TCEQ) regularly assesses surface-water and groundwater quality trends throughout the state. In a 2005 report, the TCEQ and the Texas State Soil and Water Conservation Board reported that 306 water bodies throughout Texas had impaired water quality.[30] In streams where the water quality is too poor to support fishing and swimming, the TCEQ carries out total maximum daily load (TMDL) studies to develop strategies to reduce pollution. As of January 2008, TMDL studies were completed in 26 stream segments throughout Texas, and pollution reduction policy was being implemented for these watersheds.

Groundwater

The TWDB recognizes nine major aquifer systems, which are groundwater reservoirs that provide large amounts of water over large areas (Figure 2.4), and 21 minor aquifers, which provide large amounts of water to small areas or small amounts of water over large areas. In addition, numerous other minor groundwater sources are used on a local basis. Combined, these sources provide approximately 60% of the total water use in the state.

Only in the last century did Texas begin to develop and pump groundwater supplies on a large-scale. In 1940, groundwater provided less than 1 million acre-feet of water annually. Since the drought of the 1950s, annual groundwater use has averaged around 10 million acre-feet. The volume of groundwater pumped and used varies significantly from region to region. On a statewide basis, agricultural irrigation accounts for 79% of total groundwater use, roughly 82% of which (6 million acre-feet annually) occurs in the High Plains.[31]

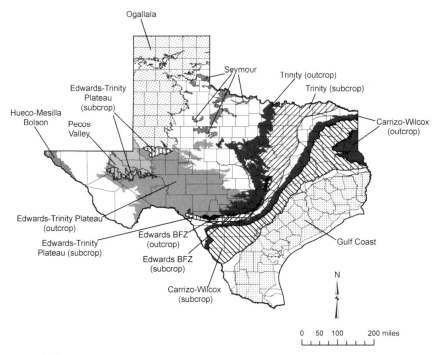

Figure 2.4 *Major aquifers*

Source: Texas Water Development Board, *Water for Texas 2007*, 177.

The amount of groundwater that is available for use under current groundwater conservation district permits and with existing infrastructure totals roughly 8.5 million acre-feet.[32] By the year 2060, groundwater supplies under these conditions are projected to fall by 32% to 5.8 million acre-feet, primarily as a result of continued depletion of the Ogallala Aquifer on the High Plains and reduced supply from the Gulf Coast Aquifer because of coastal subsidence districts' mandatory reductions in pumping to reduce land surface subsidence. Land subsidence occurs when groundwater pumping causes parts of a landscape to permanently sink, thus increasing the risk of flooding and damage to property, improvements, roads, utilities, and water and wastewater infrastructure.

One consequence of widespread groundwater use is that the depth to groundwater has significantly increased in several regions of the state.[33] The greatest declines in groundwater levels (more than 1,000 feet) have historically occurred in the north Texas region around Dallas. Yet some water levels have generally stabilized where users have converted to surface-water sources. Water level declines in the Gulf Coast Aquifer in Harris, Galveston, and Fort Bend Counties have previously caused significant land surface subsidence issues. Although water demands continue to escalate in this high-growth area of the state, groundwater pumping is now being managed to reduce subsidence damage. Elsewhere, ongoing water table declines averaging approximately 1 foot per year are occurring over parts of the High Plains in the Texas Panhandle.

Throughout the state, water quality challenges occur in several aquifers.[34] The depletion of the freshwater portion of some aquifers is resulting in increased concentration of natural (and in some cases, artificial) chemical constituents. Increases in salinity, total dissolved solids, arsenic, radionuclides, and nitrate are some of the more critical problems facing some groundwater users. In addition, common anthropogenic sources of pollution include oil and gas exploration, petrochemical production, and failing on-site wastewater treatment systems. The Texas Groundwater Protection Committee (TGPC) documented 4,729 groundwater contamination cases during the 2008 calendar year.[35] The most common contaminants (65% of documented cases) include gasoline, diesel, and other petroleum products. Among the less common contaminants are organic compounds, pesticides, creosote constituents, solvents, heavy metals, and sodium chloride.

Many communities in Texas have benefited from securing alternative water sources. An example of what can go wrong when a region does not develop multiple water sources can be found in the Houston to Galveston area along the upper Gulf Coast. For many years, municipal and industrial water suppliers depended only on groundwater sources. The pumping was so great that land subsided throughout much of the area, dropping the elevation of the land surface by as much as 12 feet and, in some areas near the coastline, to below sea level. This subsidence threatened highways, buildings, and other infrastructure and increased the risk of flood damage. Although these events cannot be reversed, in the late 1970s, the Texas legislature took the extraordinary step of creating the Harris–Galveston Subsidence District to curtail additional subsidence in the region. District policies, along with those of the Fort Bend Subsidence District, have significantly reduced groundwater pumping as the area converted to new surface-water sources.

Water Source Disparity

Although Texas has significant surface-water and groundwater resources, water supplies are not equally dispersed across the state. Over time, Texans have regionally adapted to specific water supply sources. In the eastern quadrant of the state, generally from Dallas to Houston, surface-water reservoirs have been constructed to take advantage of abundant rainfall; in the west, groundwater is the more widely used supply source.

The western part of the state has several reservoirs, but the more arid conditions there prompt reliance on groundwater sources, especially in the High Plains, Trans-Pecos, and Big Bend regions. For example, agricultural irrigators in the High Plains exclusively rely on groundwater from the Ogallala Aquifer. As that nonrenewable resource continues to be mined, groundwater users in the region have come to understand that the aquifer eventually will be exhausted. To prolong the economic viability of the Ogallala Aquifer, the High Plains community has created groundwater conservation districts with the goal of promoting agricultural and municipal water conservation through education programs and regulations.

In contrast, farmers along the southwestern border from El Paso to Brownsville rely almost exclusively on surface water from the Rio Grande, the international river that separates Texas from Mexico. There the challenge is to adapt to shortages that result from periodic below-average runoff in the river's headwaters in northern New Mexico and southern Colorado, as well as from occasionally diminished flows in tributaries originating in Mexico. In general, agricultural producers are more vulnerable to shortfalls, largely because they do not value alternative water supplies as highly as do urban water users.

The idea that some regions of Texas are dependent on specific sources of water is not limited to agriculture. Numerous municipalities faced with growing populations are wondering how they will address additional water demand. San Antonio provides an example of a city that has historically relied on a single groundwater resource—the Edwards Aquifer—and now is considering alternative conjunctive use sources. In addition to being the single source supply for San Antonio, the Edwards Aquifer also provides the only discharge to protect endangered and threatened species that live in Comal and San Marcos Springs. Water supply utilities that service the San Antonio area are evaluating potential new supply sources, including importation of groundwater and surface water from distant sources, desalination of more locally available brackish groundwater, and water and wastewater reuse. Currently they are satisfying growing demand by relying on water market operations in which Edwards Aquifer water rights are purchased from the irrigation sector. Further discussion pertaining to water markets in Texas can be found in Chapter 4.

NONTRADITIONAL WATER SUPPLY STRATEGIES

Where traditional water sources are in short supply or unavailable, Texans are turning an eye to alternative technologies. To cope with anticipated water resource shortages, the TWDB has identified a suite of nontraditional water development and management strategies aimed at reducing consumption or creating additional water supplies, or both. These strategies include municipal and agricultural conservation, desalination of brackish groundwater and coastal seawater, water and wastewater reuse, and voluntary land stewardship.

Texas is increasingly turning to conservation as a cornerstone effort of meeting projected water deficits. Simply educating water users and improving awareness of the importance of conservation may play a key role. To provide guidance in achieving this goal, the TWDB formed the Water Conservation Implementation Task Force and charged the group with creating a set of recommendations for achieving municipal, industrial, and agricultural water savings. The results of this group's efforts were published by the TWDB in 2004 as the *Water Conservation Best Management Practices Guide*.[36] In total, water conservation efforts are projected to reduce use by approximately 2 million acre-feet annually by the year 2060.[37]

Municipal water conservation strategies are anticipated to result in about 617,000 acre-feet of water savings per year by 2060[38] and include the following principles:

- aggressive plumbing fixture replacement programs (including requirements to install water-efficient showerheads, faucets, and toilets in new construction);
- water-efficient landscape codes;
- water loss and leak detection programs;
- education and public awareness programs;
- rainwater harvesting; and
- changes in water rate structures.

An old but proven method of water conservation—storing precipitation with rainfall capture systems and cisterns—is making a comeback in many parts of Texas. Recently, the Texas Cooperative Extension[39] and the Texas Water Development Board[40] have published manuals with information on how to design, build, and manage rainfall-harvesting systems. The TWDB report notes that more than 400 full-scale rainwater capture systems have been installed in central Texas and suggest that this practice is one strategy to address increasing water demands in the region.

Agricultural water conservation can be strengthened by the widespread use of new technologies (e.g., more efficient drip and low-flow irrigation systems) and improved management (e.g., better monitoring the volume of water that is applied and timing irrigation applications to meet crop water demands). Using these methods, TWDB estimates that irrigation strategies will reduce water use by nearly 1.4 million acre-feet annually by 2060.[41]

The desalination of seawater from the Gulf of Mexico and brackish groundwater from aquifers is finding greater appeal as a water supply alternative. Although modern advances in microfiltration have increased the efficiency of the process, the economics of desalination are, however, still somewhat unproven, especially in light of its energy intensity. Desalination facilities currently are permitted and operating in 49 counties throughout inland areas of Texas, and planners suggest that by 2060, desalination can provide more than 313,000 acre-feet of water annually.[42]

The Texas legislature has authorized the funding of both seawater and brackish groundwater pilot projects to demonstrate the effectiveness of this new technology. The governor's office has called for implementing the state's first large-scale seawater desalination project. Three promising sites near the coast at Brownsville, Corpus Christi, and Freeport have been evaluated, and a pilot plant study is currently under way at the Brownsville site.[43]

Another recent initiative emphasizing brackish groundwater desalination is intent on developing this relatively untapped source of water supply.[44] Large volumes of brackish quality groundwater exist throughout the state that can be desalinated to produce usable quality water.[45] A significant deterrent to the desalination process, however, is the safe and cost-effective disposal of the highly saline concentrate that is generated from the process. The city of El Paso recently constructed the largest inland desalination plant in the world. When fully utilized, this plant will be capable of desalinating brackish groundwater from the Hueco Bolson Aquifer to produce 27.5 million gallons of fresh water per day.[46] A more detailed discussion on desalination as an advanced water supply technology is addressed in Chapter 10.

Perhaps nothing better illustrates the shifts in water supplies than water and wastewater reuse. Until the 1980s, sewage was a substance nobody wanted. Now, as wastewater is treated to higher standards, areas with water deficiencies throughout the state are eying treated and recycled water and wastewater as a viable strategy to resolve water shortages. In other words, what was once a waste product may become coveted. The TWDB projects that approximately 1.3 million acre-feet of water per year, or about 14% of new water supplies, will be made available through reuse management strategies by 2060.[47]

Voluntary land stewardship is emerging as an important water management strategy in Texas. Studies suggest a relationship between the condition of a watershed and the quantity and quality of its surface water and groundwater.[48] Brush management is seen as a key component of good land stewardship. A number of studies have been carried out to determine the hydrologic gains that may occur by selectively replacing nuisance brush species with native grasses and short-rooted plants that may use less water. A 2005 report by Wilcox and others suggests that brush control may be more likely to increase streamflows in regions that receive an average of at least 17 inches of rainfall per year.[49]

HISTORY OF WATER PLANNING AND MANAGEMENT

Following what many consider to be the worst drought on record—the statewide drought that lasted throughout much of the 1950s and afflicted all but 10 of Texas's 254 counties—state legislators created the Texas Water Development Board (TWDB) in 1957. The goal was to "take action to protect Texas communities from future drought and to ensure that the state's water supplies were both dependable and adequate enough to sustain and promote future economic development."[50]

The legislature mandated that the TWDB initiate a statewide water-planning process for addressing these issues. Since the first of these *Water for Texas* plans was approved in 1961, the TWDB has created six additional statewide water plans, with the most recent completed in 2007. The next state water plan is expected to be produced in 2012. Generally speaking, most of the early plans underscored an alleged need to build dams to meet anticipated demands for water; for example, 45 new major reservoirs were recommended in the initial plan. A paradigm shift can be sensed in the 1997 plan, which for the first time recommended water conservation and reuse as viable strategies. Yet this plan continued prior traditions by promoting more dams and infrastructural solutions. Recent planning efforts have been more inclusive by including measures to increase water use efficiency, reusing existing supplies, and developing new technologies.

Similar to the 1950s drought, the 1996 drought was one of the triggering mechanisms that led the Texas legislature to rethink how best to achieve statewide water planning. In 1997, the legislature passed Senate Bill 1, a comprehensive law that established a new decentralized water-planning process led by local and regional decisionmakers (stakeholders) in 16 geographic regions throughout the state.[51] The 2002 *Water for Texas* plan was the first to incorporate water management strategies developed by these regional planning groups.

To resolve anticipated water supply shortages, regional water-planning groups identified and evaluated more than 4,500 water management strategies for the 2007 *Water for Texas* plan, which were expected to generate roughly 9 million acre-feet of water annually by 2060.[52] In addition, the 2007 plan contained recommendations for 19 major and minor unique future reservoir sites and 15 river and stream segments of unique ecological value.[53] Since Senate Bill 1 was enacted, any water project that is proposed in any region of Texas must be consistent with recommendations in regional and state water plans if it is to obtain state funding. Moreover, any entity that proposes to increase water supplies in a manner that is not consistent with the state plan must take actions to amend the current regional and state plans or seek a waiver of this requirement.

Although numerous strategies to resolve water shortfalls have been identified, it is unlikely that additional interbasin transfers, which typically move water to more arid parts of the state, will be a solution under current laws. Although nearly 200 interbasin transfers have occurred since 1968, only a few have taken place since 1997. The lack of recent interbasin transfers can be largely attributed to provisions of the 1997 Texas Senate Bill 1, which require exhaustive permits and public hearings before transfers can be approved.

CONCLUSIONS

In coming years, Texans will be further challenged by the water demands of a population that is projected to double over the next 50 years. When considering this challenge, it is essential to recognize how the diverse natural features of the state contribute to a wide spectrum of water-related issues ranging from droughts and water shortages to flooding. Just the mere size of the state manifests in varying hydrologic environments. Climatic and topographic differences alone justify dissimilar resource management approaches. Rather than view water issues through the lens of political boundaries, it is more prudent to consider the unique physical and hydrologic obstacles that occur within specific watersheds and ecosystems.

A comparison of the western and eastern halves of the state serves to demonstrate the diverse nature of water issues relevant to each region. The reduced rainfall of west Texas, as little as 6 inches annually in El Paso, provides insufficient runoff to maintain consistent flow in surface streams and only sporadic opportunities to recharge subsurface aquifers. In this region, evaporation exceeds precipitation, motivating conservation actions that generally result in the lowest per capita water use in the state. The reality of west Texas is that water is a finite resource that, in the absence of significant human adaptation, will one day be exhausted in some locales.

In contrast, east Texas typically receives annual amounts of rainfall (as much as 66 inches annually in Beaumont) that provide sufficient flows to rivers, fill surface-water reservoirs, and recharge groundwater aquifers. In many instances, water resource problems facing east Texas are more likely to be overcome by management decisions: Will new dams be built? Will water be transferred from one river basin to another? Will surface-water storage and distribution facilities be built to lessen dependence on groundwater?

Changes in water use in certain areas of the state are providing a measure of relief to some and concern to others. In areas like the Texas High Plains, most water use has involved groundwater pumping for agricultural irrigation. In many of these regions, projected water shortages are being balanced in large part by anticipated declines in irrigated acreage. Groundwater from some of the larger aquifers in the state, such as the Ogallala, Edwards, and Carrizo-Wilcox, is being targeted for exportation to urban centers both near and far. Water exportation has become a very controversial topic statewide.

Texas is currently addressing region-specific water resources challenges by engaging in a revised form of planning in which citizens of 16 designated regions talk over the issues and explore potential solutions. Since 1997, the state legislature has mandated that each region develop its own planning efforts to balance expected shortages through identifying new water sources, acquiring existing water supplies, increasing water use efficiency, considering water and wastewater reuse, and encouraging conservation to reduce water use. The TWDB combines the 16 regional plans into a state water plan that is updated every five years.

What do the state water plans show? In the municipal, industrial, and agricultural sectors, water use efficiency is increasing. Innovative farmers are using efficient irrigation systems that reduce water losses and are calculating the specific amount of irrigation to apply each day based on local soil moisture and weather conditions. Conservation in urban areas is also improving with both voluntary and mandatory measures being enacted. Most new homes and buildings are now being installed with water-conserving showerheads and toilets, xeric landscapes with drought-hardy native plants are replacing thirsty lawns, and an increasing number of homes are using captured rainfall as a primary or secondary water supply.

Some of the solutions to Texas water predicaments lie in the development and use of new technologies and innovative management solutions. For example, desalination is now being field-tested as a way to convert saline seawater for beneficial use; waters of the Gulf of Mexico are being tapped as potential water supplies for Corpus Christi and Brownsville. Following the lead of El Paso, many inland areas are exploring the potential of brackish groundwater desalination. On another front, advances in technology and management are enabling communities such as San Antonio and El Paso to treat and reuse water and wastewater to irrigate landscapes and power industries.

In the future, some sectors of Texas may find it difficult to maintain sufficient water supplies at accustomed prices to support current uses. Some farmers may find it unprofitable to irrigate. Some rapidly urbanizing areas may have to develop new and often more expensive sources of supply. As water concerns escalate, Texans will be encouraged to consider new alternatives—reuse, desalination, conservation, pricing structures, rainfall capture—that until now may have been unacceptable. In the end, those solutions may make the most sense of all.

NOTES

1. Wikipedia, Comparison between U.S. States and Countries by GDP (PPP), http://en.wikipedia.org/wiki/Comparison_between_U.S._states_and_countries_by_GDP_ (PPP) (accessed June 10, 2010).

2. Bureau of Economic Geology, *Physiographic Map of Texas* (Austin, TX: University of Texas at Austin, 1996).

3. T. Larkin and G. Bomar, *The Climatic Atlas of Texas*, Technical Report 192 (Austin, TX: Texas Water Development Board, 1983).

4. Texas Water Development Board (TWDB), *Water for Texas 2007* (Austin, TX: TWDB, 2007), 132.

5. Texas State Historical Association, The Handbook of Texas Online, http://www.tshaonline.org/handbook/online/articles/DD/ybd1.html (accessed June 10, 2010).

6. R. Riggio, G. Bomar, and T. Larkin, *Texas Drought: Its Recent History*, Technical Report (Austin, TX: Texas Department of Water Resources, 1987).

7. TWDB, *Water for Texas 2007*, 297.

8. Ibid., 136, 298.

9. B. Billingsley, *An Assessment of Municipal Drought Contingency Planning in Texas*, Public Administration Program Technical Report No. 62 (San Marcos, TX: Texas State University, 2002).

10. D. Wilhite, Government Response to Drought in the United States with Particular Reference to the Great Plains, *Journal of Climate and Applied Meteorology* 22, vol. 1 (1983): 40–50.

11. TWDB, *Water for Texas 2007*, 297.

12. Ibid., 252.

13. Ibid., 120.

14. Ibid., 120–21.

15. Ibid., 122.

16. Ibid., 2.

17. Ibid., 246.

18. Ibid., 278.

19. L. McKinney, The State of Rivers, *Texas Parks and Wildlife* (July 2004): 21–25.

20. L. McKinney, The State of Springs, *Texas Parks and Wildlife* (July 2005): 24–29.

21. C. Contreras, *Thirty Years of Investigating Fish and Wildlife Kills and Pollution in Texas*, Technical Report 2003-001 (Austin, TX: Texas Parks and Wildlife Department, 2003).

22. L. McKinney, The State of Bays, *Texas Parks and Wildlife* (July 2003): 21–28.

23. G. Brune, *The Springs of Texas* (1981; repr., College Station, TX: Texas A&M University Press, 2002).

24. G. Mallard, K. Dickson, T. Hardy, C. Hubbs, D. Maidment, J. Martin, P. McDowell, B. Richter, G. Wilkerson, K. Winemiller, and D. Woolheiser, *The Science of Instream Flows: A Review of the Texas Instream Flow Program* (Washington, DC: National Academies Press, 2006).

25. TWDB, *Water for Texas 2007*, 138.

26. J. Smith, M. Baeck, J. Morrison, and P. Rees, Catastrophic Rainfall and Flooding in Texas, *Journal of Hydrometeorology* 1 (1999): 5–25.

27. TWDB, *Water for Texas 2007*, 142.

28. Ibid., 172–73.

29. Ibid.

30. Texas Commission on Environmental Quality (TCEQ) and Texas State Soil and Water Conservation Board (TSSWCB), *Managing Nonpoint Source Pollution in Texas* (Temple, TX: TCEQ and TSSWCB, 2005).

31. TWDB, *Water for Texas 2007*, 176.

32. Ibid., 234.

33. Ibid., 223–24.

34. Ibid., 224–30.

35. Texas Groundwater Protection Committee (TGPC), *Joint Groundwater Monitoring and Contamination Report, 2008*, Report SFR–056/08 (Austin, TX: TGPC, 2009).

36. TWDB, *Water Conservation Implementation Task Force: Water Conservation Best Management Practices Guide*, Report 362 (Austin, TX: TWDB, 2004).

37. TWDB, *Water for Texas 2007*, 260.

38. Ibid., 262.

39. R. Persyn, D. Porter, and V. Silvy, *Rainwater Harvesting*, Handbook B-6153 (College Station, TX: Texas Cooperative Extension, 2004).

40. H. Krishna, *Texas Manual on Rainwater Harvesting*, Technical Report (Austin, TX: TWDB, 2005).

41. TWDB, *Water for Texas 2007*, 263.

42. Ibid., 270.

43. Ibid., 303.

44. Ibid., 302–03.

45. LBG-Guyton Associates, *Brackish Groundwater Manual for Texas Regional Water Planning Groups*, (Austin, TX: Texas Water Development Board, 2003).

46. TWDB, *Water for Texas 2007*, 302.

47. Ibid., 269.

48. Ibid., 272.

49. B. Wilcox, W. Dugas, K. Owens, D. Ueckert, and C. Hart, *Shrub Control and Water Yield on Texas Rangelands: Current State of Knowledge*, Technical Report 05-01 (College Station, TX: Texas Agricultural Experiment Station, 2005).

50. TWDB, *Water for Texas 2007*, 110.

51. Ibid., 116.

52. Ibid., 256.

53. Ibid., 267.

CHAPTER 3

Texas Water Law and Organizations

Ronald Kaiser

Texas water law reflects a confluence of geology, history, drought, and politics. This confluence has resulted from an evolutionary rather than revolutionary adaptation process that contributes to the state's legislative uniqueness. In many ways, Texas law is out of sync with the hydrologic interconnectedness of water. It is based on the three geologic containers for water: (1) surface water in lakes, rivers, and streams; (2) diffused (drainage) surface water; and (3) percolating groundwater. A different set of laws governs the allocation system and ownership in each container. Basically, surface water is state owned and allocated under the prior appropriation doctrine, groundwater is privately owned and allocated under the capture rule as modified by local groundwater conservation districts, and diffused surface water is privately owned so long as it remains in a diffused state. However, when diffused surface water enters a natural watercourse, it becomes state property subject to the prior appropriation doctrine.[1]

SURFACE-WATER LAW

At various times in its water history, Texas has followed Spanish and Mexican civil law, the English doctrine of riparian water rights, and the western doctrine of appropriative rights (see Table 3.1).[2] Although these legal systems added richness to the Texas heritage, they have generated conflicts.

During its years as a republic (from 1836 to 1840), Texas followed the civil laws of Mexico and Spain. Water rights associated with land grants made by Spain, Mexico, and the Republic of Texas are recognized and protected under Texas law. The Texas Supreme Court has ruled that if the original Spanish, Mexican, or Republic of Texas land grant did not specifically mention water rights, they cannot be implied from the laws of these three sovereigns.[3] Only those water rights

Table 3.1 *Evolution of Texas surface-water laws*

Sovereign	Dates	Water rights regime
Spain	1600–1821	Spanish civil law
Mexico	1821–1835	Mexican civil law
Republic of Texas	1836–1839	Civil law
	1840–1845	Riparian law
State of Texas	1845–1888	Riparian law
	1889–1912	Prior appropriation in west Texas/riparian law
	1913–1966	Mixed prior appropriation and riparian law
	1967–present	Prior appropriation statewide

expressly included in the grant are recognized and enforced.[4] This is contrary to the English riparian doctrine, which grants water rights to riparian land even if the land grant did not specifically mention water rights.

In 1840, Texas abolished the civil law and adopted English riparian doctrine, which it followed until 1895.[5] Land grants made by the republic and state of Texas during this 1840–1895 time frame included riparian water rights. These water rights remain with the property and continue in full force even if the water is not used. In contrast with the prior appropriation doctrine, a riparian water right is not lost if the water is not used.

Droughts are part of Texas history, and they usually have precipitated significant changes in water law. A drought cycle in the late 1880s and early 1890s threatened the developing agricultural economy and revealed weaknesses in the riparian doctrine as it applied to the arid portions of Texas. Mindful of the water law experiences of western states, the Texas legislature incrementally adopted the prior appropriation doctrine.[6] In adopting appropriation principles, the legislature preserved water rights granted under civil and riparian law, which created a conflicted water rights system. This conflicted system presented few problems when water was available to satisfy all users. During the drought of the 1950s, conflicts arose as to what law applied. The most celebrated cases involved the Rio Grande.[7] Water rights claims based on civil law, riparian law, and the prior appropriation system exceeded the amount of water available in the river, compelling the state to file suit to have a court determine the efficacy of these competing water rights claims. The litigation dragged on for 13 years, involved about 3,000 parties, and generated an estimated $10 million in court costs and attorney fees.[8] This case illustrated that after 78 years of expensive legislative and judicial attempts to reconcile the two systems, another approach was needed.

Finally, in 1967, the Texas legislature merged the civil and riparian water law regimes into the prior appropriation system with the enactment of the Water Rights Adjudication Act.[9] The act required anyone claiming a water right based on civil law, riparian law, or prior appropriation to file an application with the Texas Water Commission (now the Texas Commission on Environmental Quality) to convert his or her water claim to a prior appropriation right.[10] Landowners had a three-year time period to file their claims; if they failed to do so, they lost their preexisting water rights. Once all these claims were filed, the commission began a river basin adjudication process, completing it in the 1990s. This adjudication process resulted in the granting of more than 6,000 water rights permits. Today anyone seeking a new permit to use surface water must comply with the appropriation permit procedures outlined in the Texas Water Code.

PRIOR APPROPRIATION, TEXAS STYLE

State-owned surface water is broadly defined to include the ordinary flow, underflow, and tides of every flowing natural watercourse in the state (a watercourse has a definite bed and banks). Storm water and floodwater found within natural lakes, rivers, and streams are also state waters,[11] as are springs that form headwaters of natural streams.[12] Anyone seeking to use state surface water must obtain a permit from the Texas Commission on Environmental Quality (TCEQ).[13]

Seniority Rule

Seniority is the linchpin of the prior appropriation doctrine. The principle of "first in time, first in right" determines the allocation of water.[14] The priority date is established when the complete application is filed with the commission. When ample water exists, the seniority rule seldom matters, but when shortages occur, seniority determines who gets the available water.

Quantified Amount of Water

Under the appropriation system, a permit holder is entitled to a measured flow, or volume, of water. This provision, along with the priority rule, provides an incentive for senior appropriators to invest in diversion by assuring them of a stable water supply. The quantity is not absolutely guaranteed but is limited to the amount of water beneficially used.[15]

Beneficial Use

Definitions of beneficial use have changed over time. Texas has always considered domestic, municipal, agricultural, and industrial uses to be beneficial, but it has added other uses to the list and now considers water used for recreation, parks, and games preserves beneficial as well.[16] The state has only recently recognized the importance of environmental flows as another beneficial use of water.[17]

Transferability

Because it is a vested property right, a water permit is transferable to other users or uses. This feature allows for marketing of water rights. Transfers involve either selling a water permit or entering into a contract to sell the water.[18] This is an important legal distinction. The sale of a water permit conveys all of the rights the permit holder has to the water, whereas in a water contract, the owner retains the permit and contracts only for the transfer of the water.[19] A contract does not transfer a property right and is governed by contract law rather than property law. Water contracts can be for varied time periods, ranging from yearly leases up to a 50-year time frame. Typically, contracts are used to sell water to wholesale and retail customers.

While important legal and practical distinctions exist among transfer types, the TCEQ must approve transfers involving a change in the place and purpose or amount of water use; point, method, and rate of diversion; or location at which surplus water is returned to the stream.[20] Generally, transfers involving only a change in ownership, a minimal change in water use, or no adverse harm to other water users are the least complex and are granted without providing notice to other water users and without a formal hearing.[21] Transfers that negatively affect other water rights holders or involve a substantial change in the place, purpose, or time of use require greater review, including public notice and a public hearing. Public notice and hearing requirements add to transaction costs, and they may have a negative impact on water marketing. For an expanded discussion of water transfers and marketing, see Chapter 4.

Cancellation and Loss of Water Rights

"Use it or lose it" is a condition of Texas law. Appropriative water rights may be lost if they are abandoned or are not fully used for a 10-year period.[22] Cancellation proceedings may be instituted at any time a water right has not been fully used. Even though cancellation is theoretically possible, the Texas legislature has expressed its disfavor with this approach by establishing a cumbersome and expensive process that must be followed before a right can be canceled. Consequently, the TCEQ has not canceled a single water right in the last 25 years.

Interbasin Transfers and the Junior Rights Rule

Texas law has long permitted interbasin transfers while at the same time protecting the rights of water holders in the basin of origin. To date, more than 100 interbasin transfers have been authorized, mostly in northeast Texas and along the Gulf Coast. These transfers allowed for water marketing provided there was no significant injury to rights holders in the basin of origin. However, in 1997, the Texas legislature amended the law to discourage interbasin transfers. It did this by changing the priority date for transfers through the junior rights rule.[23] The rule mandates that any water right transferred from the basin of origin loses its seniority and becomes junior to other rights in the originating basin. This rule reduces the reliability of a water right during drought and effectively precludes interbasin water marketing.

One unintended consequence of the junior rights rule is that it has forced a number of cities to turn to groundwater to replace surface-water resources.[24] Because most of the groundwater resources are in rural areas, the junior rights rule has exacerbated rural and urban tensions over water developments, transfers, and markets.

Acquiring a Water Right

A water right can be obtained by either acquiring an existing permit or seeking a new one. One advantage of buying an existing permit is that the purchaser gets the priority date. This has importance during drought, when seniority determines water allocation. As previously outlined, acquisition of an existing permit requires approval from the TCEQ.

A litany of complex conditions must be satisfied by the applicant in order to acquire a new water permit. Before the commission can award a new permit, it must find that all of the following conditions are met:

- unappropriated water is available;[25]
- water will be beneficially used;[26]
- existing water rights will not be impaired;[27]
- the proposed use is not detrimental to the public welfare;[28]
- the proposed use is consistent with a water supply need in the regional water plan;[29] and
- reasonable diligence will be used to avoid waste and achieve conservation.[30]

In addition, the TCEQ must assess the effects, if any, of the issuance of the permit on bays and estuaries, existing instream uses, fish and wildlife habitat, and water quality.[31] Each of these findings represents a potential point of dispute in a hearing. As long as positive findings are made in all of these categories, the commission grants the application and issues the permit.

A water permit holder does not have title to the water, but only state permission to use and enjoy it. This permit is a vested property right that entitles the appropriator to certain protection against termination, loss, or infringement.[32] In addition to the regular permit, the commission has the option of granting more restrictive permits, such as seasonal, term, temporary, emergency, or bed and banks permits.[33]

DIFFUSED SURFACE-WATER LAW

Diffused surface water differs from state water and groundwater in its geologic characteristics. This water is the variant rainfall runoff flowing over the land surface. While in the diffused state, it is the property of the landowner and may be captured and used without a state permit.

Two issues frequently arise regarding diffused surface water. The first involves whether the water is diffused and subject to capture and retention by the landowner, and the second relates to landowner liability for damages resulting in a change in the natural geologic drainage pattern.

Ownership and retention issues are resolved by determining whether the runoff has reached a natural watercourse. Texas courts have defined a natural watercourse as having well-defined bed and banks and a flow of water from a permanent source.[34] Runoff that flows in rivulets into gullies, ravines, swales, or drainage ways but exists only intermittently during storms is generally considered diffused surface water. Conversely, once the runoff enters a natural watercourse, it is transformed from private into public ownership. Waters that overflow the normal bed and banks of a watercourse and onto the floodplain do not become diffused waters. Thus, landowners cannot capture this water without a state permit. Texas courts have concluded that water cresting over the bank of a river onto the floodplain remains state water.

Landowner liability for damages from change in the natural drainage pattern can arise in two ways. In the first scenario, an upper landowner develops land so that the natural drainage pattern is changed from its original course or the flow increases significantly. In the second, a lower landowner changes the natural drainage pattern, causing water to back up on or flood the upper landowner's property. The consequence of this natural flow rule is that landowners must build detention ponds if they have plans to develop their property that will cause drainage patterns to be altered in a way that will change the direction or volume of natural flow. These detention ponds must retain drainage water and release it so as to mimic the natural environment.

GROUNDWATER LAW

Groundwater receives different legal treatment than does surface water. Whereas surface water is state-owned, groundwater is the private property of the landowner. Texas follows the judicially established rule of capture in determining the amount of water that a landowner may pump from an aquifer. From legal and practical perspectives, the rule of capture is simple and straightforward: landowners have the legal right to capture and pump unlimited quantities of water from beneath their land, regardless of the effect on neighboring wells. A landowner is constrained only by the size of his or her pump, and hence the capture rule has been called the "rule of the biggest pump."[35] In a practical sense, the surface owner does not own the water, but only has a right to pump and capture whatever water is available. Additionally, neighboring landowners have this same right. Two widely cited Texas Supreme Court cases have affirmed and outlined the general parameters of this law.[36] A handful of other appellate court cases have acknowledged and followed the principles of the capture rule.[37]

Landowner Rights under the Capture Rule

Under the capture rule, a landowner has a "bundle of rights," which include the following:

- a right to drill a well on the property to any depth and size;
- ownership of the water captured and brought to the surface;
- a right to make nonwasteful use of the water;
- a right to sell or lease the water;
- a right to export water beyond boundaries of land or of the aquifer; and
- a right to reserve the groundwater when the land is sold.

Under the capture rule, groundwater access rights can be freely purchased and sold. No restrictions are made on the form of the transaction or the intended purpose or place of use for the water. Although economists have argued that establishing private property rights to a specified amount of water is necessary to foster a genuine market, this has not constrained transfers under the capture rule. The Texas practice has been to engage in land-based transactions where landowners have the right to access productive groundwater bodies. The following cases illustrate some of these creative transactions by public agencies:

- The city of Amarillo purchased rights to pump from 72,000 acres of land in Roberts County and will build a pipeline to transport the water to the city.
- The city of El Paso purchased rights to pump from 76,000 acres of land in Hudspeth, Valentine, and Van Horn Counties and is negotiating for pumping rights on an additional 25,000 acres.
- The Canadian River Municipal Water Authority purchased rights to pump from 43,000 acres of land in order to supply water to 11 cities in the Texas Panhandle.[38]

Private interests can also develop various business arrangements for the purposes of marketing groundwater. Among the more common forms are landowner partnerships, cooperatives, and private corporations. These private relationships provide for quantifying the amount of water to be produced, monitoring pumping, and transferring the water to the purchaser, thus overcoming property rights concerns. Some examples include the following contemporary cases:

- Mesa Water, a landowner partnership originated by T. Boone Pickens, has amassed 150,000 acres of land in Roberts County and is seeking a purchaser for the groundwater underlying this land.[39]
- Brazos Valley Water Alliance, a landowner cooperative, has accumulated 133,000 acres of land in Brazos, Robertson, Burleson, and Milam Counties and is seeking a purchaser for the groundwater.[40]
- Carrizo-Wilcox Water Alliance (formerly Metropolitan Water Corp.) has acquired rights to pump from about 55,000 acre-feet of water in Burleson, Lee, and Milam Counties and is seeking to build a pipeline to furnish water to a customer.[41]

Such private arrangements may engender institutional innovation. Historically, the public sector has dominated in the acquisition and development of municipal water supplies under a revenue-neutral approach, meaning that financing structures used for this purpose did not include a profit concept. These more recent private arrangements predicated on profit may precipitate new public policy.

Judicial Limitations to the Capture Rule

On occasion, Texas courts have found the results from the rule harsh and have fashioned certain exceptions to its general application. Courts have ruled that the capture rule does not apply when there is malicious pumping with the intent to harm neighbors,[42] negligent pumping that causes land subsidence,[43] or waste.

In theory, these three exceptions seem to be major constraints to landowner abuse, yet as applied by Texas courts, they have served as only minimal limitations on exploitation. For example, in *City of Corpus Christi v. Pleasanton*,[44] the Texas Supreme Court adopted the malicious pumping rule but refused to find waste in the transportation of groundwater some 100 miles through a surface watercourse, even though three-fourths of the original supply was lost in transit as a result of evaporation and seepage. Correspondingly, in the *Friendswood Development Corp. v. Smith-Southwest Industries, Inc*,[45] the court held that landowners could recover for subsidence losses caused by negligent pumping of groundwater but could not recover if their wells went dry.

Statutory Limitations

The Texas legislature also has shown some discomfort with the capture rule and has modified it with two legislative provisions: groundwater may not be subject to the capture rule where a groundwater conservation district limits pumping[46] or when it is from the underflow of a river.[47] These limitations illustrate that the legislature believes that groundwater is subject to reasonable regulation to moderate resource degradation. The Texas Supreme Court has agreed in ruling that privately held groundwater rights are subject to reasonable regulation by the state.[48]

Groundwater Conservation Districts

Groundwater conservation districts (GCDs) were legislatively authorized in 1949, and the first district was formally established in 1951.[49] Over the next 35 years, only 22 districts had formed throughout the state.[50] That slow growth rate has dramatically changed. Over the last two decades, more than 69 districts have been established. As of 2010, there were 96 districts either in operation or proceeding through the establishment process.[51] More than half of these 96 districts encompass only a single county.

District advocates contend that GCDs are locally controlled and are best suited to consider local needs in developing their management plans.[52] Critics counter that problems of self-interest, limited funding, local politics, and the self-limiting nature of these districts prevent meaningful management and protection of groundwater resources.[53] Concerns have been raised regarding the number of districts, the motives for creating them, the lack of conformity between district and aquifer boundaries, and the lack of integration and coordination between districts and regional water-planning groups. But despite the criticism of groundwater districts, legislative sentiment remains strong that groundwater should be managed locally.[54]

Legislative powers granted to GCDs reflect a mix of public policy objectives. Districts can function as planning agencies, regulators, and water providers.[55] In exercising these powers, GCDs can influence land development and local economies. The following points briefly highlight some of the mandated and optional duties and powers of GCDs:

Mandated Duties of GCDs

- develop and adopt a management plan;[56]
- adopt rules to implement the plan;[57]
- set goals to achieve one or more of the following: control subsidence, prevent waste of groundwater or degradation of water quality, or conserve, preserve, protect, and recharge the groundwater reservoir;[58]
- register and require permits for wells;[59]
- keep records on wells and the production and use of groundwater;[60]
- adopt governance rules and establish administrative and financial procedures;[61] and
- hold regular meetings.[62]

Optional Duties of GCDs

- regulate the spacing of and production from wells;[63]
- set rules to control land subsidence, prevent degradation of water quality, and prevent waste of groundwater;[64]
- buy, sell, transport, and distribute surface water and groundwater;[65]
- acquire land by purchase or eminent domain;[66]
- provide public educational materials and programs;[67]
- require wells to be capped or plugged;
- require export permits for water transported outside the boundaries of the district;[68]
- establish export fees;[69] and
- enforce rules by injunction and set reasonable civil penalties, not to exceed $10,000 per day per violation, to ensure compliance with district rules.[70]

GCD Impact on Water Marketing. Groundwater conservation districts facilitate groundwater marketing by solidifying the private property rights of landowners to a quantified amount of water, and they restrict markets by limiting transfers or imposing fees on them. A GCD can facilitate water transfers by establishing, through a permit system, a landowner right to a quantified amount of water. Therefore, a GCD regulation can facilitate a water market by establishing, through the permit, a property right to the water.[71]

GCDs can impede groundwater transfers by limiting the amount of water that a landowner can export and by imposing export fees on the water transferred.[72] Although a district may not absolutely prohibit water exportation, it may limit the amount of water that can be marketed and exported. One way districts can impede transfers is by intruding into the affairs of another governmental entity by inquiring as to their "need" for water.[73] GCDs can limit private property rights and water transfers by finding that an importing area does not need the water and use this as a basis for limiting exports.

Another district limitation that constrains water transfers is the exportation fee that may be imposed for water transported outside district boundaries.

Interestingly, water transferred within the boundaries of a district is not subject to these fees. These export fees can limit transfers by increasing transaction costs to the point where the transfer is no longer economically feasible.

The Edwards Aquifer Authority GCD. The Edwards Aquifer Authority (EAA) is a unique GCD that is responsible for managing and protecting the Edwards Aquifer, one of the major groundwater systems in Texas, serving nearly 2 million people in the San Antonio area.[74] In addition to being the primary water source for municipal, industrial, and agricultural users, the Edwards supports a unique ecosystem of aquatic life, including several threatened and endangered species.[75] The EAA differs from other GCDs in that it cannot issue pumping permits beyond a legislatively mandated numerical pumping limit, in order to ensure a continuous minimum springflow at Comal and San Marcos Springs. These springs support threatened and endangered species inhabiting the region. For a more detailed discussion of the Edwards Aquifer and its unique institutions, see Chapter 5.

SPRINGFLOW AND THE LAW

It is generally accepted that once groundwater bubbles to the surface in a watercourse, it becomes state-owned and subject to allocation under the prior appropriation system. However, if the groundwater is captured before flowing to the surface, then the capture rule applies. This means that springs, with all their economic, ecological, and social values, receive scant legal protection under the capture rule.[76] Except for the Edwards Aquifer Authority, neither the capture rule nor GCDs explicitly protects springflows. In fact, capture rule principles can directly affect springflows without consequence for the pumper. In *Pecos County Water Control & Improvement Dist. No. 1 v. Williams*,[77] the court allowed irrigators to overpump the aquifer and dry up Comanche Springs. The court ruled that groundwater became state-owned surface water only when it emerged from the ground. Before it emerged from the ground, the defendant could use any amount of groundwater regardless of the impact on surface-water users.

The proximity of a water well and pipe to a spring does not matter, so long as the well pumps the water before it flows to the surface. In *Denis v. Kickapoo Land Co.*,[78] the court held that a well sunk into the underground cavern just beneath the spring was capturing groundwater, and thus the well owner was not liable for a reduction in springflow.

WATER AGENCIES

Federal, state and local water agencies abound in Texas. They are used to achieve numerous purposes associated with water allocation, development, distribution, financing, management, planning, and protection.

Federal Agencies

In contrast to many western states, where there is a significant federal estate, nearly 98 percent of Texas land is privately owned. Consequently, the federal land footprint in Texas is small, as is the federal water footprint. Federal landholdings in Texas resulted from negotiated acquisitions rather than reservation from federal lands.[79] For this reason, the federal reserved water rights doctrine has little applicability in the state. This has contributed to a smaller federal water presence in Texas than in any other western state.[80]

The U.S. Army Corps of Engineers (USACE) and the Bureau of Reclamation (BOR) are the only federal agencies to play a role in major water development projects in the state. However, they do not have water rights associated with their facilities; storage rights at these facilities are held by state and local agencies. Conflicts over water allocation seldom occur between these two federal agencies and the state. The BOR is subject to a mandate to comply with state water laws.[81] The USACE does not have this mandate, but it has often obtained state and local sponsors for its projects in Texas.

U.S. Army Corps of Engineers.   The U.S. Army Corps of Engineers, under the Department of Defense, has a major navigation, flood control, water supply, recreation, and wetland regulatory presence in the state. It maintains about 423 miles of the Gulf Intracoastal Waterway from Brownsville to Beaumont. In conjunction with navigation, the USACE operates locks and dams on most of the 15 major river basins in the state, including those on the Trinity, Brazos, and Colorado Rivers.

In addition to its navigation responsibilities, the USACE operates 31 flood control and water supply reservoirs providing nearly 40 percent of the state's water supply.[82] The water in these reservoirs is owned by Texas and allocated under prior appropriation rules to state, local, and private entities. As a recreation agency, the USACE maintains some 345 recreation areas at 30 reservoirs.[83]

U.S. Bureau of Reclamation.   The U.S. Bureau of Reclamation is best known for the more than 600 dams and reservoirs it constructed in the 17 western states. Texas became a reclamation state in 1906, when Congress specifically included the state in the provisions of the Reclamation Act. However, the BOR does not have a major presence in the state. Of the 211 major dams and reservoirs in Texas, the BOR operates only 5: the dams at Lake Meredith near Amarillo, Twin Buttes near San Angelo, Choke Canyon near Corpus Christi, and Palmetto Bend (Lake Texana) west of Houston, as well as Mansfield Dam (Lake Travis) near Austin.

State Agencies

Water management agencies have undergone significant evolutionary change over the last 25 years. This evolution has not always been without conflict. Many of the water changes resulted from hostile interagency competition, communication lapses, alienation of politically powerful stakeholder groups, or a need to improve fiscal efficiency through consolidation. Today four state agencies are involved in water planning, development, quality protection, and allocation. Generally, regulatory activities related to water quality and the administration of the surface water rights are handled by the Commission on Environmental Quality; planning and financial assistance for local water projects by the Water Development Board; fish and wildlife interests by the Parks and Wildlife Department; and oil, gas, and mining activities involving the use of water by the Railroad Commission.

Texas Commission on Environmental Quality. The Texas Commission on Environmental Quality (TCEQ) is the environmental protection agency for Texas. The agency has broad regulatory powers over water, air, and waste pollution programs established under both state and federal legislation. A three-member governing body of commissioners, appointed by the governor, establishes policy and has supervisory responsibility over programs assigned to the TCEQ. An executive director, appointed by the commission, is the agency's chief administrative officer responsible for enforcing TCEQ rules, regulations, and policies.[84]

The TCEQ is the successor agency of the Texas Water Commission (TWC). All of the statutory powers of the TWC are now vested in the TCEQ. As the primary water protection and regulatory agency in the state, the TCEQ has jurisdiction to do the following:

• administer and allocate surface-water rights under the prior appropriation doctrine;
• administer the state's water quality program;
• regulate public drinking-water systems;
• regulate dam construction, maintenance, and removal;
• administer the oil and hazardous spill program and license hazardous waste disposal facilities;
• regulate water well drillers
• administer the National Flood Insurance Program;
• certify wastewater treatment operators;
• regulate water and sewer rates; and
• define priority groundwater management areas.[85]

Texas Water Development Board. The Texas Water Development Board (TWDB) is the planning and financing agency for state and local water projects. Unlike most state agencies, which are established by legislation, the TWDB was established by a constitutional amendment.[86] Following are the duties of the TWDB:

• coordinate and fund regional planning groups and regional plans;
• consolidate regional plans into a state water plan;
• collect and maintain water data;
• administer loan and grant programs to local governments for water supply and water quality purposes;
• develop groundwater availability models and projections for the state's major and minor aquifers;

- administer the environmental Water Trust; and
- assist in financing state and local water projects.[87]

Texas Parks and Wildlife Department.   Responsibility for the development and protection of water-based recreation and wildlife falls to the Texas Parks and Wildlife Department (TPWD). This department provides recreational facilities and enforces water safety as well as fish and game laws. The TPWD evaluates and recommends changes to water right applications and modifications to ensure protection of aquatic and riparian ecosystems; and acquires water rights to protect instream flows and freshwater inflows to Texas bays and estuaries.[88]

Texas Railroad Commission.   The Texas Railroad Commission (TRRC) enforces regulations to protect groundwater and surface water from wastes generated by oil and gas production. The TRRC also regulates surface mining of coal and lignite, and uranium and water resources associated with that mining activity.[89] If surface water is used in oil and gas production, the driller or producer must comply with TCEQ guidelines. For groundwater, TRCC rules apply to exploration and production, and GCD rules may apply to some elements of production.

Local Water Entities

Texas has a bewildering array of local public and private water institutions that form the state's water industry and are responsible for water supply and management. Although state agencies have broad statewide responsibilities, most water management is undertaken by regional and local water entities.[90] These entities, generally referred to as districts, have powers to tax, issue bonds, acquire land, and construct and operate water infrastructure and facilities.

Water districts were first authorized in 1904 through the adoption of Article III, Section 52 of the Texas Constitution, granting the legislature the power to pass laws establishing these types of districts.[91] Districts initially were used to develop the state's agricultural resources, but they have evolved to provide water and wastewater services for urban and rural areas. Over the years, the legislature has been especially creative in establishing a byzantine mix of water districts, as shown in Table 3.2. The subsequent sections highlight some of the more prominent districts.

Municipal Utility Districts.   In order to provide water services in suburban areas not covered by a city, the legislature authorized the creation of municipal utility districts (MUDs).[92] In addition to water, MUDs are also authorized to provide sewer, drainage, park, fire protection, street lighting, recreation facilities, solid waste disposal, and roads.

MUDs are political subdivisions with the authority to issue tax-exempt municipal bonds to finance and construct facilities and infrastructure. They are governed by publicly elected boards of directors and operate under governance rules for local public agencies. A unique feature of these districts is that they

Table 3.2 *Water districts in Texas*

District type	Active	Inactive	Dissolved
Municipal utility	650	290	449
Water control and improvement	173	44	500
Groundwater conservation	92	1	19
Freshwater supply	54	12	89
Drainage	43	5	58
Special utility	38	16	2
Levee improvement	31	15	78
River authority	31	0	1
Navigation	24	2	8
Irrigation	24	1	3
Water improvement	17	0	40
Municipal management	14	27	4
Others	50	29	53
Total	1,241	442	1,304

Source: Texas House Committee on Natural Resources, *Interim Report 2006, A Report to the House of Representatives*, 80th Texas Leg., November 29, 2006.

operate with few of the taxing and bond indebtedness restrictions imposed on cities and counties. The TCEQ retains oversight over the MUD in that audit, engineering, and feasibility requirements must be satisfied in order for the MUD to issue tax-exempt bonds and other financial instruments.

A MUD can be created in one of two ways: by a petition process through the TCEQ or through a special act of the legislature.[93] Under either process, a city's consent is required if a MUD is established within its corporate or extraterritorial limits. The city has the legal option to annex the MUD, acquire all of its assets, and assume its debt. For MUDs located outside the extraterritorial jurisdiction of any city, the county commissioner's court is entitled to review the establishment petition. However, the county does not have veto authority over the establishment of the MUD.

Although they can be formed anywhere in the state, MUDs are the primary financing tool for developers in the greater Houston area. More than 520 of the 650 active MUDs are located in the Houston-Harris County area. The MUDs have flourished because of the inability, or unwillingness, of the city of Houston to extend services to these rapidly growing areas outside its corporate limits.

Water Control and Improvement Districts. Water districts have undergone substantial changes. Initially established for irrigation, water districts have changed names as they have been given additional responsibilities. Many irrigation districts morphed into water improvement districts and then into water control and improvement districts (WCIDs). Today WCIDs have authority to perform the following duties:

- operate dams and reservoirs;
- supply raw untreated and treated water to municipal, industrial, and other customers;

- operate wastewater (sewer) facilities;
- control drainage and floods;
- supply water for irrigation and maintain irrigation channels;
- provide hydroelectric power; and
- undertake navigation projects.[94]

In addition to these water-related activities, WCIDs may acquire and maintain parks, campgrounds, and other recreation facilities and provide for their programming, security and maintenance. In order to finance water and recreation activities, districts have the power to levy operation and maintenance taxes and may issue revenue bonds.

The Tarrant Regional Water District (TRWD), located in the Dallas–FortWorth metroplex, exemplifies this type of district. TRWD is one of the largest raw water suppliers in the state of Texas, providing water to more than 1.6 million people. Some of its wholesale customers include the cities of Fort Worth, Arlington, and Mansfield, plus the Trinity River Authority. Operations span a 10-county area and include maintaining four water supply and flood control reservoirs, 150 miles of pipeline used for water transport, 27 miles of floodway levees, more than 40 miles of Trinity River trails, and a 260-acre wetland water reuse project aimed at increasing future water supplies for the area.[95]

River Authorities.   River authorities are multicounty districts established by a special act of the legislature for the purpose of managing surface waters in a particular basin. Consequently, each river authority is as unique as the basin it manages.[96] River authorities own and operate 22 major water supply reservoirs, with a combined storage capacity of more than 10 million acre-feet. Authorities are major water wholesalers, as an estimated 43 percent of the surface water is controlled by river authorities. They in turn contract to sell this water to public and private water suppliers (see Table 3.3).[97]

Although some 31 districts include "river" in their names, only 15 of those hold significant water rights.[98] River authorities generally do not have taxing powers, but they can do all of the following:

- operate dams and reservoirs;
- supply raw untreated and treated water to municipal, industrial, and other customers;
- operate wastewater (sewer) facilities;
- control drainage and floods;
- supply water for irrigation and maintain irrigation channels; and
- provide hydroelectric power.

River authorities also may acquire and maintain parks, campgrounds, and other recreation facilities and provide for their programming, security, and maintenance. They also have powers of eminent domain, meaning they may acquire land and facilities by negotiated purchase or through condemnation.

Table **3.3** *District water rights by river basin*

River basin	District amount (million acre-feet)	Basin total (million acre-feet)	Percentage of basin total
Brazos	1.16	5.86	20
Canadian	0.16	0.16	99
Colorado	4.32	6.82	63
Cypress	0.26	0.52	50
Guadalupe	3.76	6.23	60
Lavaca	0.08	0.16	50
Neches	1.52	4.36	35
Nueces	0.05	0.55	10
Red	0.33	0.67	50
Rio Grande	2.83	6.98	40
Sabine	1.35	2.28	59
San Antonio	0.08	0.18	46
San Jacinto	0.15	0.63	24
Sulphur	0.10	0.50	20
Trinity	1.34	4.53	30
Total	17.49	40.43	43

The Lower Colorado River Authority (LCRA), established in 1934 by special enabling legislation, is an example of a river authority.[99] Its boundaries cover 10 counties through which the Texas Colorado River flows. LCRA controls floods, supplies low-cost electricity for central Texas, manages water supplies, develops water and wastewater utilities, provides public parks, and supports community and economic development in many Texas counties. It is the largest river authority in the state based on employees, of which it has more than 1,800.[100] LCRA has no taxing authority and operates solely on utility revenues and fees generated from supplying energy, water, and community services. Its largest source of revenue comes from selling wholesale electricity to more than 40 retail utilities, including cities and electric cooperatives that serve more than 1 million people in 53 counties.

LCRA owns and operates six dams, controlling flooding and providing water supplies for cities, farmers, and industries along a 600-mile stretch of the Texas Colorado River between San Saba and the Gulf Coast. It provides water to retail and wholesale customers in 11 counties and more than 30 communities, including the city of Austin. It also operates 20 regional and local wastewater treatment facilities and an environmental laboratory. The authority enforces ordinances that control illegal dumps, regulates on-site sewage systems, and reduces the impact of major new construction along and near the lakes. It is also a major park and recreation agency, as it owns about 16,000 acres of recreational lands along the Highland Lakes and Colorado River, operating more than 40 parks, a natural science center, and nature preserves. The Highland Lakes attract more than 1 million visitors a year to its parks, campgrounds, and recreation facilities.

INTERSTATE COMPACTS AND TREATY WATERS

Texas shares river water with New Mexico, Colorado, Oklahoma, Louisiana, and the Republic of Mexico. These sharing arrangements are governed by two international treaties and five interstate compacts. The Rio Grande is apportioned by two treaties and a tristate compact, the Red River by a four-state compact, the Canadian River by a tristate compact, and the Pecos and Sabine Rivers by bistate compacts. These legal agreements generally apportion the waters among the states, and they established institutions to supervise and enforce agreement provisions. Waters allocated to Texas by these treaties and compacts are then apportioned according to the prior appropriation doctrine. For further information on treaties and compacts, see Chapter 8.

CONCLUSIONS

Most of the surface water in Texas has been fully allocated under prior appropriation rules, and little conflict arises over these rules. Legal and legislative skirmishes will continue to flare over designation of reservoir sites, water reuse, removal of transfer barriers, interpretation of contract language in water agreements, and compliance with conservation mandates. However, two significant legal issues—providing water for environmental flows and the junior rights limitations on interbasin transfers—have yet to be resolved and are likely to provoke significant changes to existing law.

Environmental flow issues exist because most of the surface water in the state was allocated without regard to the importance of maintaining a certain level of environmental flow in rivers and to estuaries. A number of state agencies and stakeholder groups have been going through a process to determine how much water is necessary for these flows. Once completed, the real challenge will involve crafting legal rules to reallocate water from existing uses and users to these unaccounted-for environmental flow needs.

Removing barriers to the interbasin transfers of surface water is the second major legal issue for Texas. Historically, Texas law did not impose significant barriers to interbasin transfers.[101] In Senate Bill 1, passed during the 75th Texas Legislature Session (1997), the Texas Water Code was amended to expand the requirements for interbasin transfer, the most onerous being the junior rights provision. Resolving the junior rights transfer requirement presents a prickly challenge for Texas stakeholders, policymakers, and politicians, as it pits water interests in rural east Texas against urban water demands of the Dallas and Houston metropolitan areas. Resolving this type of geographic supply-versus-demand conflict likely will ultimately be reconciled in favor of the group with the greater support of the Texas legislature.

Refining the parameters of groundwater regulation will continue to be the dominant legal and political issue in the state. Whereas surface water is owned and managed by the state, groundwater is privately owned and allocated based on the capture rule. This rule may be modified by a locally established groundwater conservation district (GCD). Most of these districts were formed over the last two decades to deal with the harsh consequences of the capture rule, in the hope that local management would solve regional and statewide groundwater issues. Proponents contend that districts are locally controlled and thus best suited to consider local needs in developing their management plans. Critics counter that groundwater districts do not have the funds to develop meaningful regulations, have not prevented overpumping and mining of groundwater, have made decisions based not on sound science but on the politics of the moment, do not want to regulate their neighbors, and do not encourage meaningful water conservation. As districts begin the task of regulating pumping by private landowners, litigation doubtless will follow.

NOTES

1. *Hoefs v. Short*, 273 S.W.2d 785 (1925); *Turner v. Big Lake Oil Co.*, 96 S.W.2d 221 (1936); *In re Water Rights of Lower Guadalupe River Segment, Guadalupe Basin and a Portion of the Lavaca-Guadalupe Coastal Basin (Green Lake)*, 730 S.W.2d 64 (Tex. App.–Corpus Christi, 1987, writ ref'd n.r.e.); *Domel v. City of Georgetown*, 6 S.W.3d 349 (Tex. App.–Austin, 1999, pet. denied).

2. For a history of Texas water law, see W. Hutchins, *The Law of Texas Water Rights* (Austin, TX: Department of Water Resources, 1961); Hans W. Baade, The Historical Background of Texas Water Law: A Tribute to Jack Pope, *St. Mary's Law Journal* 18 (1986): 1.

3. *State v. Valmont Plantations*, 355 S.W.2d 502 (Tex. 1962).

4. If water rights under a Spanish or Mexican land grant are expressly stated in the grant, then the landowner has a right to the natural flow from the stream. In *San Antonio River Authority v. Lewis*, 363 S.W.2d 44 (Tex. 1964), the court held that Lewis had a vested right to water that included the right to the flow in a natural channel. The court held that the authority must compensate the landowner for damages caused by its having moved the river channel some 200 feet away from its natural course.

5. The common law was adopted by the Republic of Texas in 1840 (see Tex Rev. Civ. Stat. art. 1, Rep. Tex Laws 3 [1840]). The 1895 date was established by the 1913 Irrigation Act as the last on which the state granted water rights with its land patents.

6. With the passage of laws in 1889, 1985, and 1913, Texas moved to appropriation. See ch. 88, 1889 Tex. Gen. Laws 100, 9 H. Gammel, *Laws of Texas* 1128 (1898); ch. 21, 1895 Tex. Gen. Laws 21, 10 H. Gammel, *Laws of Texas* 751 (1898); and ch. 171, 1913 Tex. Gen. Laws 211.

7. *State v. Hidalgo County*, WCID No 18, 443 S.W.2d 728 (Tex. Civ. App.–Corpus Christi, 1969, writ ref'd n.r.e.).

8. D. Caroom and P. Elliot, Water Rights Adjudication—Texas Style. *Texas Bar Journal* 44 (November 1981): 1183–84.

9. Texas Water Code, §§ 11.301–11.341.

10. This conversion was approved by the Texas Supreme Court. In re *Adjudication of Water Rights of Upper Guadalupe Segment*, 642 S.W.2d 438 (Tex. 1982).

11. Ibid.

12. *Fleming v. Davis*, 37 Tex. 173 (1872).

13. Permits are often termed certificates of adjudication.

14. Texas Water Code, § 11.027.

15. Ibid., § 11.025.

16. Ibid., § 11.023.

17. Ibid., § 11.0237.

18. Ronald Kaiser, Texas Water Marketing in the Next Millennium: A Conceptual and Legal Analysis, *Texas Tech Law Review* 27 (1996): 196–203.

19. An example of this type of transaction involved the 2002 sale by the Garwood Irrigation Company of its state-issued water rights permit to the Lower Colorado River Authority. This water rights permit was for 133,000 acre-feet of water per year and had a 1900 priority date, the most senior water right in the Colorado River Basin. The Lower Colorado River Authority purchased the 1900 permit for an estimated $75 million.

20. Texas Water Code, § 11.122.

21. Ibid., § 11.122(b).

22. Ibid., §§ 11.171–11.186.

23. Ibid., § 11.085(s).

24. Scott Parks, Water Investors Eye Liquid Assets: Demand Creates a Market for Aquifer Right in Texas, *Dallas Morning News*, May 21, 2000.

25. This provision applies only when a person is seeking a new water right. In cases where an existing water right is transferred, the only right that can be conveyed is that associated with the original permit. The commission can attach additional conditions on water transfers that may have the effect of modifying some of the original permit conditions. As Texas rivers become more completely appropriated, the availability of new permits will continue to decline. According to the commission, 12 of the 15 major river basins in Texas are fully appropriated, and no water is available for a new appropriation. TCEQ, Water Availability Models, http://www.tceq.state.tx.us/permitting/water_supply/water_rights/wam.html#statewide (accessed December 11, 2008).

26. An important step in perfecting an appropriation is the intention to put water or its actual application to a beneficial use. Once an appropriator puts water to a beneficial use, the right is perfected and becomes a vested property right. The statutory list of beneficial uses of water include domestic, municipal, industrial, irrigation, mining, hydroelectric power, navigation, recreation and pleasure, stock raising, public parks, and game preserves. Texas Water Code, § 11.023(a).

27. Commission staff evaluate possible impacts on downstream water rights, and if the proposed appropriation would impair water availability for existing downstream rights, the commission may impose diversion point and timing restrictions in the permit. A typical restriction would limit diversions under the new permit when the flow of the stream at the diversion point is less than a specific number of cubic feet per second.

28. The commission must consider the environmental, social, and economic impacts of any proposed appropriation. As part of this process, the commission assesses the effects of habitat mitigation, water quality, estuarine impacts, and instream uses in considering a permit to store, take, or divert state waters.

29. A water right for municipal purpose may not be issued by the commission in a region that does not have an approved regional water plan. However, the commission has the discretion to waive this requirement. Texas Water Code, § 11.134 (c).

30. Ibid., § 11.1271. Applicants for new or amended permits must submit a conservation plan demonstrating that reasonable diligence will be used to avoid waste. Conservation considerations include those practices, techniques, and technologies that will reduce the consumption of water, reduce the loss or waste of water, improve efficiency in the use of water, or increase the recycling and reuse of water so that a water supply is made available for future or alternative uses. All conservation plans must include quantified targets for water savings. In addition, municipal and industrial users that use more than 1,000 acre-feet per year and irrigators that use more than 10,000 acre-feet per year must submit and implement a water conservation plan consistent with the regional water plan.

31. Ibid., § 11.147.

32. For loss of rights through cancellation proceedings, see ibid., § 11.172. Interfering with or impairing a water right without improper authority is unlawful (ibid., §§ 11.081–11.083).

33. A right to use state water can be acquired by a new appropriation, purchase and transfer of an existing water right within a river basin, purchase or transfer of an existing water right from another basin, or contractual sale or purchase (lease of water). In a contractual sale, the owner of a water right sells water but retains ownership of the water right.

34. *Hoefs v. Short*, 278S.W. 785 (1925).

35. Ronald Kaiser, *Handbook of Texas Water Law* (Austin, TX: Texas Water Law Institute, 1986).

36. *Houston & T.C. Ry Co. v. East*, 81 S.W. 279 (Tex. 1904), established the capture rule and *Sipriano v. Great Spring Waters of America, Inc.*, 1 SW2d 75 (Tex. 1999), reaffirmed the rule.

37. See *Barshop v. Medina Co. Underground Water Conservation District*, 925 S.W. 2d 618 (Tex. 1996); *Denis v. Kickapoo Land Co.*, 771 S.W.2d 235 (Tex. App.–Austin, 1989, writ denied); i, 643 S.W.2d 681 (1983); i, 558 S.W.2d 75 (Tex. Civ. App.–Houston [14th Dist.], 1977, writ ref'd); *Friendswood Dev. Co. v. Smith-Southwest Indus., Inc.*, 576 S.W.2d 21 (Tex. 1978); *City of Corpus Christi v. Pleasanton*, 276 S.W.2d 798 (Tex. 1955); *Pecos County W.C.I.D. No. 1 v. Williams*, 271 S.W.2d 503 (Tex. Civ. App.–El Paso, 1954, writ ref'd n.r.e.); *Lower Nueces Water Supply District v. City of Pleasanton*, 251 S.W.2d 777 (Tex. App.–San Antonio, 1952, no writ).

38. Environmental Defense, *A Powerful Thirst* (Austin, TX: Environmental Defense, 2004), available online at http://www.texaswatermatters.org (accessed December 12, 2008).

39. See Mesa Water Inc. website, http://www.mesawater.com (accessed December 12, 2008).

40. See Brazos Valley Water Alliance LP website, http://brazoswater.com/index.php?option=com_content&task=view&id=56&Itemid=67 (accessed June 1, 2010).

41. See Mary Alice Kaspar, Water Fight Goes to Court, *Austin Business Journal*, March 7, 2003, http://austin.bizjournals.com/austin/stories/2003/03/10/story3.html (accessed February 12, 2008).

42. *City of Corpus Christi v. Pleasanton*, 276 S.W.2d 798 (Tex. 1955).

43. *Friendswood Development Corporation v. Smith-Southwest Industries, Inc.*, 576 S.W.2d 21 (Tex. 1978).

44. *City of Corpus Christi v. Pleasanton*, 276 S.W.2d 798 (Tex. 1955).

45. *Friendswood Development Corporation v. Smith-Southwest Industries, Inc.*, 576 S.W.2d 21 (Tex. 1978).

46. Texas Water Code, ch. 36.

47. Ibid., § 11.021. Underflow is not defined by statute, but one court has held that it is that portion of the flow of a surface watercourse occurring in the sand and gravel deposits beneath the surface of the streambed that is hydrologically connected to the surface flow of the stream. *Texas Co. v. Burkett*, 296 S.W.2d 273 (Tex. 1927).

48. *Barshop v. Medina County Underground Water Conservation District*, et al., 925 S.W.2d 618 (Tex. 1996).

49. Texas Water Code, ch. 36.

50. See Kaiser, *Handbook of Texas Water Law.*

51. See TWDB, GCD Map, http://www.twdb.state.tx.us/mapping/maps/pdf/gcd_only_8x11.pdf (accessed June 1, 2010).

52. See Texas Alliance of Groundwater Districts website, http://www.texasgroundwater.org (accessed December 12, 2008).

53. See Joe Greenhill and Thomas Gee, Ownership of Ground Water in Texas: The East Case Reconsidered, *Texas Law Review* 33 (1955): 620, 629 (urging Texas courts and the state legislature to adopt rule prohibiting malicious waste of water); Corwin W. Johnson, Texas Groundwater Law: A Survey and Some Proposals, *Natural Resources Journal* 1017, no. 22 (1982):

1024 (discussing wastefulness of absolute ownership of percolating groundwater); Corwin W. Johnson, The Continuing Void, in Texas Groundwater Law: Are Concepts and Terminology to Blame? *St. Mary's Law Journal* 17 (1986): 1281, 1293 (addressing the absence of a legislative declaration of state ownership of groundwater); Eric Behrens and Matthew Dore, Rights of Landowners to Percolating Groundwater in Texas, *Texas Law Review* 32, no. 5 (1991): 185, 191 (commenting on the Texas Supreme Court's and legislature's refusal to change the rule); Jana Kinkade, Compromise and Groundwater Conservation, *State Bar of Texas Environmental Law Journal* 26 (1996): 230, 233 ("Not only has the Texas Legislature been slow to act, but the Texas courts have impeded the progress of Texas ground water law"); David Todd, Common Resources, Private Rights and Liabilities: A Case Study on Texas Groundwater Law, *Natural Resources Journal* 32 (1992): 233, 256 (criticizing the law of Texas groundwater management); Ronald Kaiser and Frank Skillern, Deep Trouble: Options for Managing the Hidden Threat of Aquifer Depletion in Texas, *Texas Tech Law Review* 32 (2001): 249.

54. Texas Water Code, § 36.0015.

55. For a discussion of the powers, duties, and limitations of GCDs, see B. Lesikar, R. Kaiser, and V. Silvy, *Questions about Groundwater Conservation Districts in Texas*, Texas Cooperative Extension, Publication B-6120, 06-02 (College Station: Agricultural Communications, 2002), available online at http://texaswater.tamu.edu/groundwater.GCD.htm (accessed June 1, 2010).

56. Texas Water Code, § 36.1071.

57. Ibid.

58. Ibid., § 36.101.

59. Ibid., § 36.113.

60. Ibid., § 36.111.

61. Ibid., § 36.154.

62. Ibid., § 36.064.

63. Ibid., § 36.116. A district may regulate production limits based on acreage or tract size, a defined number of acres assigned to an authorized well site, acre-feet of water per acre of land, or gallons per minute per well site, or managed depletion.

64. Ibid., § 36.101(a).

65. Ibid., § 36.104.

66. Ibid., § 36.105.

67. Ibid., § 36.110.

68. Ibid., § 36.064.

69. Ibid., § 36.122.

70. Ibid., § 36.102(b).

71. See Clay Landry, *A Free Market Solution to Groundwater Allocation in Texas* (Austin, TX: Texas Public Policy Foundation, 2000), suggesting that water permits are an element of a private property rights system.

72. Texas Water Code, § 36.122, deals with water exportation and generally provides that although a district cannot prevent the sale and export of water by a landowner, it can limit the amount of water that can be exported outside the district by considering the availability of water in the district and in the proposed receiving area; the projected effects of the transfer on aquifer conditions; depletion, subsidence, or existing users within the district; and whether the transfer is congruent with their district management plan and the state regional water plan.

73. Ibid., § 36.122(f)(1).

74. The 1993 Edward Aquifer Authority Act defines the Edwards Aquifer as "that portion of a belt of porous, water-bearing, predominately carbonated rocks known as the Edwards and associated limestones in the Balcones Fault Zone extending from west to east to northeast from the hydrologic division near Brackettville in Kinney County that separates underground flow toward the Comal Springs and San Marcos Springs from underground flow to the Rio Grande

Basin, through Uvalde, Media, Atascosa, Bexar, Guadalupe, and Comal counties, and in Hays County south of the hydrologic division near Kyle that separates flow toward the San Marcos River from flow to the Colorado River Basin." See the Edwards Act of June 11, 1993, ch. 626, 1993 Tex. Gen. Laws 2350 § 1,03(l), at 2351.

75. In response to the specter of federal regulation of the Edwards Aquifer, the Texas legislature made a Faustian choice and opted for state rather than federal control of the aquifer. Under the aegis of an affirmative constitutional duty to conserve Texas natural resources, the legislature created the Edwards Aquifer Authority and granted it the authority to regulate aquifer withdrawals. See the Edwards Act of June 11, 1993, ch. 626, 1993 Tex. Gen. Laws 2350; as amended by Act of May 16, 1995, 74th Leg., R.S., ch. 524, 1995 Tex. Gen. Laws 3280; Act of May 29, 1995, 74th Leg., R.S., ch. 261, 1995 Tex. Gen. Laws 2505; Act of May 6, 1999, 76th Leg., R.S., ch. 163, 1999 Tex. Gen. Laws 634; Act of May 28, 2001, 77th Leg., R.S., ch. 966, §§ 2.60–2.62, 6.01–6.05, 2001 Tex. Gen. Laws 1991, 2021–22, 2075–76; and Act of June 1, 2003, 78th Leg., R.S., ch. 1112, § 6.01(4), 2003 Tex. Gen. Laws 3188, 3193.

76. Ronald Kaiser, Who Owns the Water, *Texas Parks and Wildlife Magazine* (July 2005): 31–35.

77. 271 S.W.2d 503 (Tex. Civ. App.–El Paso, 1954, writ ref'd n.r.e.).

78. 771 S.W.2d 235 (Tex. App.–Austin, 1989, writ denied).

79. Texas entered the nation as an independent republic reserving the rights to its lands. This is in contrast to many other western states, which were established out of federally controlled lands where the federal government reserved certain lands from settlement.

80. Four federal departments—Agriculture, Defense, Interior, and Commerce—as well as the U.S. Environmental Protection Agency (EPA) and numerous agencies under them, have some involvement in Texas waters. Some dams are operated by the U.S. Army Corps of Engineers under the Department of Defense and the Bureau of Reclamation, in the Department of the Interior. Water quality falls within the purview of EPA. Involvement in the protection of water is shared by the Soil Conservation Service and the USDA Forest Service in the Department of Agriculture and by the National Park Service, U.S. Fish and Wildlife Service, and U.S. Geological Survey in the Department of the Interior. The National Oceanic and Atmospheric Administration in the Department of Commerce provides climatic and hydrologic information.

81. U.S.C. § 383 (§ 8 of the Reclamation Act).

82. Compiled from websites for the Fort Worth, Galveston, and Tulsa districts.

83. See U.S. Army Corps of Engineers, Recreation Information on Ft. Worth District Lakes, http://www.swf-wc.usace.army.mil/cgi-bin/rcshtml.pl?page=Recreation (accessed June 1, 2010).

84. Texas Water Code, § 5.221.

85. See TCEQ, About the TCEQ, http://www.tceq.state.tx.us/about (accessed December 11, 2008). The TCEQ has approximately 2,900 employees and 16 regional offices, and it had a $480.7 million operating budget for the 2007 fiscal year (including both baseline and contingency appropriations). Most of the budget (85 percent) is supported by program fees, with the remainder coming from state general and federal funds.

86. Texas Constitution, art. III, § 49-c.

87. See TWDB, About the Texas Water Development Board, http://www.twdb.state.tx.us/about/aboutTWDBmain.asp (accessed December 7, 2008).

88. See the Texas Parks and Wildlife Department website, http://www.tpwd.state.tx.us (accessed December 7, 2008).

89. For a discussion of authority, see Railroad Commission of Texas, http://www.rrc.state.tx.us/ (accessed June 1, 2010).

90. Fear, necessity, land speculation, hydrology, droughts, floods, and even politics have been the drivers for establishing the near alphabetic array of local public and private water entities. Public water entities are frequently called "water districts," whereas private entities are termed "water corporations."

91. The 1917 conservation amendment provides the constitutional authority to enable the legislature to establish local water districts (Texas Constitution, art. XVI, § 59). An earlier amendment (art. III, § 52) was the first to authorize the formation of surface-water districts. General law districts derive their authority from an enabling that may apply to entities on a statewide basis, whereas special law districts are targeted to a specific geopolitical location. They are often termed "local laws."

92. Texas Water Code, chs. 49, 54.

93. For TCEQ creation, a petition must be filed with the executive director and signed by persons holding a majority of land value within the proposed district. The TCEQ evaluates the petition, holds a public hearing, and grants or denies the petition. Once approved, a publicly elected board of directors manages and controls all of the affairs of the MUD subject to the continuing supervision of the TCEQ. Texas Administrative Code, ch. 30, § 293.11.

94. Texas Water Code, ch. 51.

95. See Tarrant Regional Water District website, http://www.trwd.com/home.aspx (accessed December 12, 2008).

96. No standard geologic, hydrologic, or geopolitical criteria exist for a river authority, though all have the term "river authority" in their names. "River authority" implies a political entity that has jurisdiction over an entire river, with broad powers and duties to manage the river's water resources. The Texas legislature has deviated from the basinwide concept in the creation of river authorities and has even combined rivers into a single authority. In fact, not everyone agrees on the number of river authorities. Some suggest that "river authority" attached to an organization's name does not qualify it as one, nor does the lack of that designation preclude an organization from functioning like an authority. See Jayson K. Harper and Ronald C. Griffin, *Regional Management of Water Resources: River Authorities in Texas*, MP-1666 (College Station, TX: Texas Agricultural Experiment Station, 1988); Texas Senate Interim Committee on Natural Resources, *Missions and Roles of Texas River Authorities*, Interim Report to the 77th Leg., November 2000.

97. Kaiser, Texas Water Marketing.

98. Because no legislatively set criteria exist for defining a river authority, their reported numbers vary depending on the author or legislative committee studying them.

99. Information on LCRA is taken from the authority's website, http://www.lcra.org/about/index.html (accessed December 7, 2008).

100. The LCRA is governed by a board of directors appointed individually by the governor for six-year terms. The board consists of 15 members.

101. Prior to passage of Senate Bill 1, the Texas Water Code, § 11.085, prohibited transfer of water from one watershed to another if it would prejudice any person or property within the watershed from which the water was taken.

CHAPTER 4

Texas Water Marketing and Pricing

Ronald C. Griffin

T he combination of laws described in the previous chapter leads to a unique style of water management within Texas. Although some features of this legal system are very capable and have even been "frontierish" in responding to the demands of a changing water future, other aspects fall short of sponsoring good water stewardship. Taken together, the water marketing and pricing systems arising from these Texas laws form an instructive model.

In light of the sizable population and economic growth that continue to occur in the state (see Chapter 2), permanently locking all water into prescribed uses is undesirable. Some nimbleness in water allocation better serves social interests, so laws and organizations should be flexible enough for people and businesses to respond to their changing circumstances as weather cycles through wet and dry periods, technologies advance, and water scarcity worsens with added growth. Water consumers of all types are understandably diverse in how they use water and the values they attach to specific uses, and policy should attempt to appreciate this diversity. Addressing this goal is a comparative advantage for incentive-based policies and a tough task for regulatory policies that operate by dictating water use practices. It is also important to recognize that the environment is an important facet of these social interests.

People and businesses have difficulty making socially responsible choices when the signals they face are misleading. This chapter is about the economic signaling that occurs in Texas. To a large extent, these signals are the product of the legal system identified in the Chapter 3. A story arising here is that some Texas institutions promote good signaling and some do not. Thus, both positive and "don't-do-this" lessons can be extracted from the Texas model of water management.

Observers in other countries or even in other parts of the United States are occasionally surprised by the degree of decentralization in Texas water management. Not much detail in terms of water allocation actually is prescribed by a centralized authority, at least not until recently. That is, it is not specified by agencies, courts, or the legislature how much water either city A or irrigation district B is permanently limited to use. Even the so-called "river authorities" of Texas administer only their own water right holdings. Texas water laws have tended to grant people or their water suppliers a high degree of ownership in the water they use, and these laws also permit transactions whereby water rights, or at least rights to access water, can be exchanged for money. The details of ownership are not constant, and the rules governing trades are also evolving. Historically, a high degree of flexibility has been built into this system. The resulting water markets are quite varied, so one objective of this chapter is to inventory the market types and their properties. Not all of them are best-case role models.

Where the economic signaling performed by water trading leaves off, the signaling generated by rate-setting begins. In Texas, rates are set by the utilities, districts, and river authorities that rely on rates for revenue generation. Each such water supply organization must be independently solvent, as cross-subsidization involving other functions is discouraged.[1] The state does not set rates, as may be done in some countries, and administrative oversight of rate-setting by the state is light-handed. From the perspective of signaling water scarcity with accuracy, the performance of Texas water-pricing institutions is not as developed as it is for water marketing. However, a positive point is the evolving interface between pricing and marketing, as described later.

SIGNALING: GOOD AND BAD

Under water scarcity conditions such as those occurring in Texas, the presence of markets or prices is a half measure when society's objectives are to extract the greatest good from its water resources. That is, the mere establishment of markets and prices is an incomplete goal. The goal is to foster economically efficient markets and prices. Because of various technical problems that are widespread for water allocation, if the rules are not carefully established, traders and price setters will not arrive at terms that reliably advance economic efficiency in water use. Thus, carefully imposed market rules and pricing guidance are needed if markets and prices are to serve the public interest.

Water is an unusual economic commodity—not so much because of highly hyped human dependencies on water, but because water's flow through our environment links humans in an array of complex interrelationships. Consequently, water use and trade have many unintentional impacts. Bargains between water traders or between water suppliers and their clients are commonly accompanied by effects on third parties as a result of such things as return flows, instream flow impacts, aquifer drawdown, and saltwater intrusion, to name a few. Unless some protections or powers are then granted to third parties, marketing and pricing operations do not necessarily induce good water use stewardship. Thus, the signaling available through the economic incentives of markets and prices underachieves unless the rules are thoughtfully designed. Therein lies a warning label for those who may imagine that the promotion of markets and prices is an adequate foundation for achieving efficient water use. Fortunately, the literature of water resource economics provides a full examination of the occasional shortcoming of these policy approaches as well as the available repairs.[2] Although many of the important points of this literature cannot be revisited here because of space constraints, some of these issues arise during this chapter's review of the Texas waterscape.

MARKETING VS PRICING

A relatively sharp, practical distinction between marketing and pricing is to reserve the term "marketing" for instances in which *natural* (unprocessed, raw) water is traded among willing parties. Marketed water may be surface water or groundwater, but it must be naturally occurring water that has not been transformed in any avoidable way.[3] It has not been transported or treated by its current owner or owners. Whereas market trades of water are accompanied by financial terms that may indicate an agreed-upon price, the term "pricing" is reserved here for situations in which a water supplier is setting rates for *processed* (retail or wholesale) water provided to customers. Hence rate-setting and pricing are the same thing. The tap water used by households is a widely recognized example of priced retail water. Another example is water delivered via canals or pipelines to irrigators. Such suppliers are typically water utilities, companies, or districts.

Texas river authorities also engage in pricing. Some river authorities sell retail water to final consumers such as industries or irrigators, but they more notably price the wholesale water they sell to utilities and districts. As in the generic depiction of Figure 4.1, utilities or districts will further process (and reprice) wholesale water before it gets to final consumers, incurring more costs with each processing refinement.

Some city utilities and water districts also act as wholesalers within their regions. Indeed, around the state, myriad water supply organizations perform varied roles, undoubtedly arising from opportunities and comparative advantages during the histories of their service areas. This, in and of itself, is a success story stemming from decentralization.

Wholesalers such as river authorities commonly require that clients agree to and sign a long-term delivery contract. Contract terms specify an array of details, including financial terms such as volumetric prices and quite possibly the supplier's authority to revise prices periodically. These delivery contracts serve to stabilize the supplier's revenue and provide documentable income sources that assist suppliers in securing loans and issuing bonds. Yet these delivery contracts are traditionally weak examples of water marketing for the simple reason that water value is not captured within financial terms. Often intrinsic water value has been completely omitted in these contracts. What actually is being costed is the entirety of the supplier's expenditures in administering its functions and converting raw,

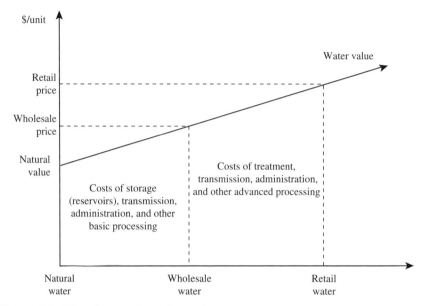

Figure 4.1 *Water's stage-dependent value*

natural water into the processed water taken by clients. Hence delivery contract pricing addresses value-adding activities such as reservoir storage rather than the base value of the resource.

Being careful with the market–price delineation helps underscore the distinct values that are embedded in any given water price. With this separation, any prices observed within a water market identify only the value of natural water—a very important signal and a geographically variable one too. Signaling is equally important at the wholesale and retail levels. Referring again to Figure 4.1, water pricing for processed water should involve both the value of natural water and the value of wages and goods used in the transformation of natural water into the stored, transported, treated, and otherwise improved water that is ultimately received by customers. Whether this is actually the case for Texas institutions is an important matter to consider.

TEXAS WATER MARKETS

Because the most significant Texas water rights are perpetual,[4] traders can employ alternative approaches in their deals. Permanent transfers of rights to new owners are called purchases or sales. Leasing refers to a onetime transfer, with the original owner retaining permanent ownership and future use of the right. Rights are divisible, so it is possible to sell or lease only part of a water right. More complex arrangements such as options are also practical. With options, the owner agrees to a multiyear deal in which the owner will continue to regularly use a water right unless a preagreed event occurs (such as a rigidly defined drought condition), at which time the option buyer can temporarily choose (opt) to use the right. With an option, the buyer may make payments to the seller at one or more of the following times: on signing the contract, regularly during the life of the agreement, or when the option is exercised. Sales, leases, and options are the most commonly recognized water-marketing instruments, but other arrangements may be devised by traders as well.

The varied platform of Texas law establishes different sets of privileges for water right owners of surface water and groundwater. Indeed, different systems of both surface-water law and groundwater law operate in distinct regions of the state.[5] The various laws, combined with different scarcity conditions across the state and the high costs of transporting water, create particularized water markets. Five individual Texas water markets are identified in Figure 4.2. Each of these marketplaces is considered separately in following sections.

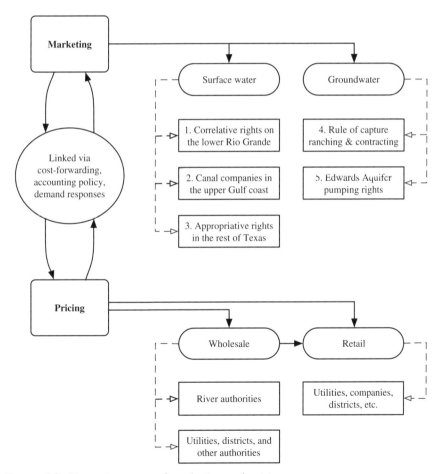

Figure 4.2 *Texas instances of marketing and pricing*

Rio Grande Valley

The most entrenched Texas water market operates in the lowest reach of the Rio Grande Valley, stretching from Amistad Reservoir to the Gulf of Mexico. This river forms the border between Texas and the country of Mexico, so international agreements are in place pertaining to the sharing of these waters. A 1944 treaty precisely states sharing ratios for the Rio Grande and its tributaries, although enforcement of these terms has been problematic in recent years (see Chapter 8 for details). The Texas market involves only those rights assigned to the state's side of the border, as no international trade of water rights is permitted in this basin.

The evolution of lower Rio Grande water law in Texas was uniquely redirected by a mid-1950s drought, which eventually led to the judicially installed system of correlative rights that has been serving the region since 1970. At that time, most water rights were assigned to the many irrigation districts of the region as well as to individual irrigators who operated their own pumping works. These correlative rights have no "first in time, first in right" seniorities as apply to surface-water rights in the rest of Texas. The only seniority respected for these Rio Grande rights applies to that of municipal rights (including domestic, residential, commercial, and industrial uses) over agricultural rights. All municipal rights have equal standing with one another, as do all agricultural rights. Both municipal and agricultural rights are quantitatively expressed and limited. In the event that stored water in the region's two-reservoir system is inadequate to satisfy all rights, agricultural water right owners share what is available after municipal rights are fully satisfied. Sharing is in proportion to water right ownership.

In spite of the severe water scarcity experienced in this region, relatively high levels of population and economic growth have been accommodated by the post-1971 water market. Cities and towns have regularly purchased agricultural water rights or entered into water contracts with irrigation districts. Because of the higher priority of municipal rights, intersectoral transfers from agricultural to municipal use employ a two-to-one conversion, achieved via a state-administered amendment procedure.[6] Thus, a theoretical agricultural right to 100 acre-feet of water becomes a firm municipal right to 50 acre-feet. Irrigation districts are also expected to obtain this amendment for any water rights that they lease or contract to municipalities.

Early in this market's history, urban utilities tended to purchase water rights from farmer owners of irrigation rights. This activity steadily concentrated the pool of untransferred irrigation rights to that owned by the area's 28 irrigation districts. These districts are nonprofit cooperatives serving memberships that use shared water conveyance facilities, mainly pumping plants and canals. Irrigation districts favor nonpermanent instruments such as leases or long-term contracts in their marketing and pricing relations with towns and cities. These irrigation districts also have employed differing rules about whether their members are allowed to lease water internally or externally (inside or outside the district).

Intrasectoral leasing of water is a common activity in the region. Farmers are more likely to lease among one another than are cities or towns. Interestingly, the greater exposure of irrigation rights to water supply fluctuations has caused irrigation lease prices to be more variable from year to year than municipal lease prices. In recent years, a combination of drought conditions and treaty failure has increased the social value of agricultural lease markets. Because of the variety of irrigation activities in the basin, ranging from pasture, cotton, and low-valued field crops to high-valued citrus and vegetables, the agricultural value of water can vary from farm to farm, so an important market accomplishment has been to accommodate the cultivation of more valuable crops, using only voluntary arrangements among farmers.

The most momentous achievement of this market has been its role in accommodating economic development. The majority of water rights now employed in municipal, commercial, and industrial uses have been acquired via this water market. Yet agriculture remains vibrant and promising, accounting for as much as 85% of current surface-water rights.[7] Although additional water development options for this arid region are occasionally discussed, no new reservoirs have been built in four decades, in spite of the challenges imposed by rapid population and economic growth. Thus, neither residents nor nonresident taxpayers have been burdened by the large expenses of further surface-water-supply investments. That is a significant achievement that likely involves many millions of dollars in averted expenditures.[8] For the most part, water reallocation has been achieved without fanfare as a result of the simple, gradual, and decentralized activity of this marketplace.

The success of this marketplace is due not only to the transferability of rights and the presence of water scarcity, but also to two unique and important features of the region: the absence of either return flow considerations or monopolies.[9] These conditions greatly improve the capability of a water market to promote social welfare. At the time of this writing, permanent water rights are transferring in the neighborhood of $2,000 per acre-foot.

Urbanization in the region has repurposed some of the irrigated lands previously served by the irrigation districts. One source of debate has been whether some of the irrigation districts' water rights should be reallocated automatically to urban utilities as farmland is converted to homes and businesses. Up until recently, irrigation districts continued to hold the water rights as their serviced acreage shrank, but they generally have been willing to enter into water supply contracts with nonagricultural utilities. The 2007 legislature acted to modify this situation with a portion of Senate Bill 3. Municipalities can now claim well-defined water rights as urbanization occurs in this region (only). This legislation establishes processes for determining the amount of transferred rights and their costs.[10] Irrigation districts receive compensation on a less-than-market-value, 68% basis. These changes do not eliminate the former market, but they modify the market's course from this point onward.

Gulf Coast Canal Companies

Sometimes water rights are bundled with the infrastructural assets of a water supply organization being acquired by another water supply organization operating in the same region. Whenever this occurs in Texas, the new owner continues to serve the clients of the purchased supplier, but ownership of water rights is transferred. Consequently, the possibility arises for the new owner to redirect water to other uses, users, and places over time.

Many public and private irrigation organizations have arisen in Texas since the late 1800s. Some subsequently failed; others merged with similar organizations; still others were purchased, becoming part of larger organizations; and some evolved new identities. At the time of their creation, the earliest of these irrigation "companies" focused on building canals and pumping plants for the delivery of river water to nonriparian (off-river) properties. At that point in Texas history, the major value of these organizations lay in their infrastructure. Surface water was far less scarce than the ability to put surface water where it was wanted. Hence the financial worth of such companies once depended on their infrastructural components. Today such acquisitions are motivated more by the value of the companies' water right holdings.

Commencing in the 1930s, the Texas legislature created a number of river authorities with extensive water management responsibilities. These authorities do not actually oversee all water management activities in their service areas, as their names suggest, but focus on their own water rights and infrastructural investments. Thus, they operate in basins or parts of a basin alongside other water right holders and water supply entities. As nonprofit organizations without the benefits of state funding, they are sensitive to revenue opportunities as well as costs. Their water right and infrastructural ownership arises largely from two sources: participation in new water supply developments (reservoirs) and purchases of canal companies. Many of the more significant reservoirs owned by these authorities were federally constructed or assisted.[11] In recent decades, the growth of river authorities' water supplies has been dominated by acquisitions. Because river authorities are averse to relinquishing control by ever selling water rights, their control over the state's water resources has steadily increased since their creation.

The primary region of canal company purchases is the eastern half of the Texas Gulf Coast prairie, which spans several river basins and generally lies within 80 miles of the coast. Here relatively flat lands, tight soils, and abundant water were especially conducive to the irrigation of water-intensive crops such as rice, as long as water could be conveyed to the right lands. Texas rice production increased sharply around 1900, and many canal systems were established in those early years.[12] The advent of river authorities came later, and their early work emphasized the development of new dams upstream in the several watersheds crossing the Gulf Coast prairie. Eventually, river authorities enlarged their operations as well as their water control by buying preexisting water supply organizations. A 1988 report lists 10 large purchases beginning in 1944, although this may not have been a complete inventory.[13] Additional acquisitions have occurred since that time. Throughout this period, the value of the new purchases has increasingly shifted to the intrinsic worth of the acquired water rights.

One of the more active suitors of canal companies has been the Lower Colorado River Authority (LCRA). This river authority operates in reaches of a basin having considerable rice acreage. The LCRA is a unique Texas river authority in that it is the only one deriving the majority of its income from electricity services.

More than 90% of its 2009 revenue ($1.3 billion) was obtained from power sales. Having entered the electricity business in the 1930s as a consequence of its hydropower capability, it purchased its first thermal plant (lignite fueled) in 1942.[14] The LCRA later built other thermal power plants and became the dominant power provider in its service area, which includes the capital city of Austin. As a substantial electricity supplier with excellent revenue-generating ability and the public sector's privilege to engage in tax-exempt borrowing, its ability to acquire canal companies has been unparalleled in the state.

By purchasing canal companies, the LCRA has extended its control of the basin's waters, established a high degree of monopolistic market power in the supply of wholesale water, eliminated the legal disputes over water that the authority traditionally experienced with these companies, and obtained the latitude to reallocate water to growing communities over time. As an example, in 1998 the LCRA purchased the privately owned Garwood Irrigation District for $75 million, ultimately gaining control of another 101,000 acre-feet of water annually while continuing to serve Garwood's irrigators. These are contentious undertakings, as neither the authority's preexisting clientele nor its new clients are 100% supportive of these changes. Yet the LCRA's unique advantages as a well-financed river authority, together with the wide range of challenges it has faced over the years, have brought it into a leadership role. For example, many of the original contractual terms and other rules the LCRA develops for its water management activities later are adopted by other Texas river authorities.

Rest-of-Texas Appropriative Rights

Outside of the lower Rio Grande, Texas surface-water law establishes transferable, appropriative (seniority-based) rights that are specified individually with details such as annual water amount, type of use, location of the diversion from a particular watercourse, rate of diversion (e.g., cubic feet per minute), and the seniority date.[15] Thousands of such rights exist in the state, and these records are centrally maintained by the Texas Commission on Environmental Quality (TCEQ). As for other water right types in the state, these permits can be exchanged in various imaginable ways, including sales, leases, and options. If all the labeled conditions of the permit are unmodified and the only difference is a new owner, then it is a simple procedure to officially register the new owner. However, the promise of water marketing as a scarcity-coping or development-assisting strategy emphasizes potential changes in the sector of use and the location of the diversion. These types of trades require much more administrative oversight because of the potential to modify streamflow patterns or interfere with the water rights exercised by other users. For these reasons, such trades require that the water right be formally amended using procedures established by the TCEQ. Specific amendment procedures are discussed below in a section reviewing administrative oversight.

In spite of the apparent possibilities, water market transactions of appropriative surface water rights are not commonplace in the state.[16] Griffin and Characklis list four reasons for the paucity of trading, some of which will change as future scarcity intensifies:[17]

- During most years, water scarcity is low in the easternmost basins. Thus, the potential for water marketing is low in these areas.
- Texas does not possess a major east-west aqueduct for transporting water to areas of greater scarcity.[18] Moreover, such a conveyance probably would have large costs relative to its benefits.
- Metering of surface-water withdrawals and other procedures required for strict water right enforcement are not uniformly conducted in the state, except for the lower Rio Grande. In the absence of careful enforcement, water users have limited motivation to forge water deals.
- The presence of river authorities in central and eastern Texas basins establishes a strong monopolistic influence that weakens the foundations for open water markets. In such instances, prospective city buyers of water are unlikely to act before a river authority has locked up a dominant proportion of water resources. Also, river authorities are quite willing to work with cities through long-term delivery contracts. Thus, not only is marketing marked by large, one-way acquisitions by river authorities, but these water wholesalers also provide a substitute contract and pricing instrument thereafter.

For the most part, these four observations are not indictments of Texas water policy, even though they limit the range of water marketing. Where water scarcity is low, marketing is not attractive, and it is even arguable that strict water right enforcement would be a wasteful application of public funds. Also, if east-to-west reallocation is not economically viable in the absence of an expensive aqueduct, then such reallocation is not publicly attractive. Finally, delivery contract mechanisms are available for achieving reallocation, as the river authority model partially suggests. In the absence of deeper study, it remains possible that such mechanisms are the best practical approaches to addressing intensified scarcity. Yet present monopolistic power raises questions about the absence of competition that might achieve better water-pricing signals or provide reference checks regarding the costs of river authority operations.

Groundwater Ranching and Contracting

Except for the Edwards Aquifer, Texas groundwater law is uniquely dominated by the rule of capture (also called the absolute ownership doctrine), as noted in Chapter 3. Because of pressing issues associated with the Edwards Aquifer, however, the rule of capture no longer applies to that single groundwater body. Also, the rising role of groundwater districts has begun to moderate the powers granted to individual landowners under a pure rule of capture.[19] Given the evidence to date, such moderations constitute a less-than-satisfactory solution to the poor performance of the rule of capture.[20]

The key element for understanding water marketing under a rule of capture is to realize that this is not a system of private property in water. It is a doctrine of private property in land and, consequently, *access to* groundwater via landownership. All the owners of land overlying an aquifer have private property in land, in that they know the boundaries of their individual landholdings and can sell or lease that land. Yet they own groundwater only as common property in Texas. That is, they possess group ownership of the aquifer's water in that (a) except for land-based transfers, no one outside the landowning group is allowed to withdraw groundwater from their aquifer; (b) individual landowners do not face limits or protections expressed in terms of quantified water entitlements; and (c) landowners cannot sell, lease, contract, or option water in the ground. Only two methods exist for transacting water within this legal framework. Neither of these procedures allows traders to advantageously use an aquifer's natural ability to move water, because ownership is landlocked.

In the first trading method, landowners may "reduce groundwater to ownership" by pumping it to the surface. Once this is done, the landowner has established title to those units of water and may transport and sell them. Legal limits to a landowner's ability to do this are largely absent in Texas. This is not a popular method of marketing groundwater because of its general inferiority to the second method. However, entrepreneurial proposals are under consideration in which landowning groups are offering to pump and deliver groundwater substantial distances using new, privately financed pipelines if the price is right.[21]

In the second method, landowners may sell, lease, or otherwise contract what they actually own: access to a groundwater body. Such exchanges can be made by various paths. As in some other areas of the western United States, land is sometimes exchanged for the purpose of transferring groundwater access. Such activity is referred to as "groundwater ranching," and the land buyer is typically a city or urban water supply organization. In western Texas, some speculative land buying has been motivated recently by the potential future worth of underground water.

As a related approach in Texas, the right to access groundwater is a separable property right, so landowners can sell land while exclusively retaining the right to pump underlying groundwater. More popularly in Texas, a landowner will retain ownership of the land while leasing the right to pump. The most common expression of these opportunities occurs when a Texas town or urban water supply organization establishes a long-term contract with a landowner. These contracts tend to be 20 years or longer (often 40 to 50), and they allow the lessee to develop groundwater from the landowner's land and transport it elsewhere. Because of deeply rooted experience with petroleum extraction contracts in Texas, such instruments are well accepted. In a groundwater setting, the lessee establishes all necessary wells, storage, and conveyance infrastructure (on the lessor's land if necessary), and the landowner receives preagreed payments during the life of the contract. The majority of these payments are usually dependent on the metered volume of water extracted, though the contract may include "take-or-pay" terms specifying a minimum quantity of water to be paid for annually regardless of whether the lessee actually extracts that much. Because of the occasional tendency to model contracts after petroleum leases, some groundwater contracts specify prices that are fixed ratios (such as 1/8) of the retail water prices charged by the contracting city utility.

Although one can herald the inventiveness of these contractual approaches for achieving water transfers to higher-valued uses, neglected third-party effects and absent regard for too-rapid depletion constitute unremedied issues. For example, if one of an aquifer's overlying landowners enters into an urban contract whereby pumping is increased, the water table will fall, with negative implications for other pumpers. Because other landowners are not compensated for their losses, society cannot be confident that such market activity promotes overall economic welfare. That is, this system is likely to be underachieving in comparison with policies available to replace the rule-of-capture doctrine. In regions where groundwater resources receive little natural recharge and the pace of depletion is a social concern, this issue is exacerbated by the neglect of water's rising social value over time.[22]

Edwards Aquifer

There are revolutionary junctures in the evolution of water institutions when society casts out the old rules and fashions new ones. A relatively recent example is that of the Edwards Aquifer, a hydrologically unique and economically significant groundwater body in central Texas. Changes to groundwater law applicable only to this aquifer occurred during the early 1990s as a result of heated controversy. These changes established an entirely new course for this groundwater-dependent region, particularly as it relates to water use limitations and the marketability of water while it is still underground. Because this aquifer is the focus of Chapter 5, attention here is confined to the pro-marketing features of recent reform.

As noted previously, the rule of capture establishes a common property rule in groundwater. In areas of groundwater scarcity, such a rule has a limited ability to motivate efficient levels of conservation. As one controls only what one pumps under the rule of capture, no one gains a title to conserved water, and the incentive to conserve is consequently slight. Under this landlocked system, the transfer of water in the ground is limited to the transfer of water access, and metering of groundwater use is unnecessary. Because actual private property in water requires that everyone's holdings be respected (with clearly understood paper "fences" concerning water and prohibitions on trespassing), the establishment of quantified water rights and use of metering are necessary, yet not sufficient, elements for achieving efficient marketing.

After lengthy adjudication procedures assigning new Edwards water rights, initiated when the 1993 Texas legislature was forced by environmental litigation to construct new water law for the aquifer, permits to withdraw specified amounts of aquifer water began to be issued in 2001.[23] During the troubled years before 1993, as well as in the years following passage of this landmark legislation, water users voiced considerable objections to legal reform. A common protest pertained to the prospective loss of the "private" property right to pump as much water as the landowner wished. However, the new law actually established original private property in water by severing the common property in water that previously was attached to private property in land under the rule of capture.

As an unusually flow-dominated aquifer with high transmissivity, the Edwards' overriding problem lies in allocation of available recharge, not in the rate of aquifer depletion over time. The recharge is highly stochastic, varying from 43,700 acre-feet in 1956 to well over 2 million acre-feet in 1992. The aquifer behaves like a natural interbasin transfer project, transporting surface water recharge from one river basin and depositing it as springflow in a more easterly river basin. Intense recharge periods sometimes fill the aquifer to capacity, whereas drought periods can threaten springflow given the high levels of pumping that now occur. It is notable that this aquifer supports considerable crop irrigation, and San Antonio, the nation's seventh-largest city, depends almost entirely on this aquifer for its water supply. Consequently, water pumpage is substantial. Some of the springs have flows sufficient to initiate rivers, and their long existence has encouraged unique (and extinguishable) species to evolve there. Hence a new legal system was needed that could foster better use of available aquifer water while guaranteeing at least threshold springflows.

The Texas approach forged by the 1993 law was to mandate metering of wells for the first time, establish quantified water rights based on documentation of past water use, deny permanent water rights to owners of wells drilled after 1993, and limit eventual aquiferwide water pumping to 400,000 acre-feet annually starting in 2008.[24] These water rights are transferable. The 1993 law also identified the adjudication procedures to be followed in seeking and approving these new water rights. A new regulatory agency was established for carrying out these missions, and the previous, less empowered management authority was terminated.

One challenge the new authority faced is the incongruity of the 400,000 acre-foot target with the legislatively guided adjudication process. Following the dictated process has resulted in permits totaling more than 550,000 acre-feet.[25] Although this outcome was expected, given the extent of overuse in 1993 and the political desire to respect the expectations of landowners and investments of well owners, more achievements were required to improve the survival probabilities of springflow-dependent species should a worst-case drought occur. Purchase and retirement of some of the newly created water rights is an available path for the management agency, which is expressly allowed to collect user fees for this purpose, but the 1993 law also required a proportional and permanent reduction in all water rights should the total exceed 400,000 acre-feet on January 1, 2008. Although the stochastic character of Edwards Aquifer recharge recommends a seniority-based rights system or some other hierarchical procedure linkable to variable recharge,[26] the 1993 law did not select such an approach.[27] In recent years, regional interests have sought means of avoiding the 400,000 acre-foot constraint in order to more fully exploit the aquifer for extractive purposes, and in 2008, they achieved some measure of political success in doing so by lifting the pumping limit to 572,000 acre-feet (see Chapter 5). Because this relaxation reduces downstream surface-water flows and is probably incongruent with endangered-species preservation, it will be interesting to witness management responses during the next extended dry weather cycle.

Various forms of water marketing have been sparked by this legal transformation. A 2004 compilation by the aquifer authority indicates that more than 1,000 water transfers had taken place involving nearly 200,000 acre-feet.[28] This is a demonstrably vigorous marketplace. These transfers include a range of unequal exchanges, such as leases, sales, and resales, so it cannot be concluded that a given percentage of the available water rights has already changed hands. Recent sales, with San Antonio as the dominant buyer, have been in the area of $5,000 to $6,000 per acre-foot.

Although formal analysis of preadjudication market activity is not available, even the promise of this marketplace is known to have initiated market transactions. Once the 1993 law was in place, providing irrigators, cities, and other agents with clear knowledge of the new rules and water right formation process, additional land transfers and contracts began to reallocate forthcoming water rights. Some speculative activity became part of these transactions, as buyers and sellers often have different expectations regarding the future worth of these water assets. Although it was not purely a water-marketing scheme because water rights were not yet operational, in the spring of 1997 the aquifer authority orchestrated an irrigation suspension program, in which $2.3 million was paid to bidding irrigators who were required to cease irrigation on participating acreage during that year.[29] Overall, these policy experiments and changes combine to make the Edwards Aquifer management the most debated Texas natural resource issue in recent memory.

A MARKETING ASSESSMENT

A report card for Texas water markets has to assign grades for at least two subjects. Passing scores for both of these subjects are required if water marketing is to be assessed positively. The first subject deals with achievements in the allocation of withdrawn water—that is, the natural water that can be diverted from streams, rivers, lakes, and aquifers. If these waters are not well allocated, then the welfare of the state's people is reduced unnecessarily. The second subject focuses on the interplay between diverted or pumped water and water that is left instream or in-ground. In this subject area, market transactions can affect both how much water stays instream for environmental, recreational, and scenic purposes and how much stays in aquifers for water table preservation, springflow protection, future use, and even land elevation support. These decisions are crucial determinants of the welfare Texans receive from their state's water resources.

With respect to the allocation of withdrawable water (the first subject), Texas water markets appear to have performed relatively well. Open markets such as those in the Lower Rio Grande Valley and the Edwards Aquifer clearly are guiding water to more valuable uses over time. Where this has been occurring, prosperity is advancing in spite of growing water scarcity. Moreover, costly water development projects have been avoided as a consequence of the options that a water market gives to growing sectors. It is in this context that the legal transformation performed for the Edwards Aquifer is especially interesting. By choosing a new legal doctrine that allows marketability, the Texas legislature was able to build considerable flexibility and resilience into the region's system. None of this implies that Texas water markets were created perfectly, but their achievements appear to be high relative to more restrictive or regulatory approaches that could have been selected.

In other ways, marketing progress with withdrawable water appears to present mixed rewards. The "big play" purchases of Texas river authorities have a greater resemblance to corporate acquisitions than to water marketing. These buyers are wholesalers, and their purchases offer the promise of long-term reallocations rather than immediate ones. As they are governmental entities with politically appointed or elected board members, reallocation can be a slow process with significant organizational and decisionmaking costs. So although river authorities can guide water to higher-valued uses through contracts with retailers and industry, it is not always a smooth, transparent, or best-case process. Also, the dominance of river authorities in eastern basins fouls the idealistic economic notion of having many buyers and sellers so that no one entity can exert control over water right prices and scarcity values can be widely visible. General water marketing requires a certain critical mass of transactions so that participants can observe customary terms (especially prices) and transaction costs can be lowered. Yet the growth of river authorities has deterred open water markets in eastern basins. From this perspective alone, further mergers and acquisitions by river authorities are arguably undesirable.

Another mixed outcome emerges from rule-of-capture groundwater contracting. On the one hand, it is admirable to witness low-value water users voluntarily lease their rights of access to those with higher water values. Without this device, surface-water imports or developments, or more costly forms of groundwater reallocation, would have been pursued; groundwater ranching or land condemnations by cities are substitute possibilities here. On the other hand, two-party agreements are suspect social instruments in a many-party hydrologic system where water tables get lowered and future demands for a depletable commodity are weakly regarded. That is, two-party trades in this legal system can and do ignore impacts external to their individual cares, implying that this type of marketing is underachieving for society at large. Indeed, it was this rule's inattention to springflow protection that caused it to be ousted for the Edwards Aquifer. This observation brings us to the second assessment area.

Although this point is oft overlooked or underappreciated, human activities are importantly dependent on leaving some water where it naturally occurs. Later chapters of this book identify some of the reasons why instream or in-ground waters are worthy of public attention and specialized policy. Some of the crucial issues involve the biological needs of freshwater species and riparian habitats; bay and estuary inflows for a variety of purposes, including reproduction activities by offshore species; the ability to assimilate pollutant loadings generated by cities and industries; and the provision of stream-based recreation. Water markets can be helpful or harmful in attempts to achieve an efficient balance between the amounts of withdrawn and preserved water in each watershed. But Texas experience suggests that water markets have been poor tools, in and of themselves, for addressing this particular problem.

In keeping with traditional Texas values pertaining to unrestricted use of personal property, water rights of all types can be used in a variety of ways. Thus, it has long been possible for a person or group to acquire water rights and then repurpose them to a new use—including leaving the water in its natural body.

Yet transactions to the environment remain rare. In those regions of the state where water markets are well established, participants are quite focused on purely diversionary applications of water. Clublike organizations or environmental groups, which serve as the most obvious champions of naturally occurring water, rarely use water markets for this purpose. The state has attempted to better enable actions of this type through the creation of a formal Texas Water Trust, but it is little used.[30] The weak contribution of water marketing to augmenting in-place water would be more tolerable if the state had a strong program addressing instream flow control, but it does not. Only in recent years has the state begun to address this issue with some seriousness (see Chapter 7). Historically, adjudication procedures granting water rights to offstream water users were not very respectful of instream flows, so Texas policy has produced a situation in which some watersheds are misbalanced in terms of diversions versus flows. Prospects for employing water marketing to correct this issue remain poor at this time unless agencies of government enter as environmental buyers.

WATER PRICING

As depicted earlier in Figure 4.1, water pricing becomes the relevant signaling instrument once natural water is improved through purposeful processing by water suppliers. The most recognized forms of processing are transportation and treatment, but other improvements are also performed. Many Texas water suppliers own water rights and perform all processing needed to deliver retail water to their clients. A clear example is the urban utility that pumps its "own" groundwater or exercises its river rights and then undertakes all activities needed to get safe water to its many customers. Such cases have no wholesale step. As with other commodities, water is not definitively wholesale or retail at any single degree of processing. For this reason, Figure 4.1 (page 52) displays a continuum of qualities, and the wholesale level is merely an intermediate stage that is concluded at a juncture where responsibilities and money change hands.

Operationally and administratively, wholesale and retail prices are similarly formed in Texas. However, this does not mean that the differences between these two prices are small, as we shall see. What it does mean is that the procedures for determining wholesale and retail prices are largely the same, and they are similarly regulated in the state.

Specific prices are not dictated by the Texas legislature or its agencies. Indeed, Texas prices, especially retail ones, vary considerably. Some for-profit (privatized) water suppliers operate in the state, but most retail water suppliers are nonprofit utilities or districts. Wholesalers are strongly dominated by the presence of public, not-for-profit organizations. Most suppliers operate with the objective of cost recovery, so they wish to resolve a rate structure that produces sufficient revenue to cover their expenditures. Although water operations are supposed to be self-funded, in the sense that revenues are the result of water bills levied on customers, it remains true that this objective is achieved unevenly across the state.[31]

Typical water consumers do not have cause to participate in water markets, so if consumers are to receive motivational notifications about water scarcity, these signals should be contained in the prices they are charged for water. This is but one of the necessary pricing details that must be satisfied if economically efficient water allocation is to be achieved across sectors, businesses, and people.[32] Therefore, if wholesalers and retailers are properly valuing their natural water inputs, perhaps with informational assistance from a water market, and if they incorporate these values in water prices, then good pricing signals can be generated. Among other things, this implies progressively higher tap water prices in more arid regions of the state, assuming that the value-adding costs of water processing are similar throughout Texas. Some evidence pertaining to this goal is assembled in the following sections.

Wholesale Water Pricing

It is difficult to say how many Texas water suppliers may be involved in wholesale operations. Major agents such as river authorities tend to be clear-cut wholesalers, though some river authorities are not and others may participate in retail water service, especially to irrigators or industries. Some urban water suppliers also sell water to neighboring suppliers, especially to suburban utilities or rural districts. Such water may require only further transportation and repricing. Some urban water suppliers sell so-called wholesale water to industries even though the receiving firm is the final consumer, but in this case the wholesale label usually means the water was less than fully treated. A state as diverse as Texas has myriad arrangements, practices, and labels in place.

To obtain some basic insights regarding the wholesale pricing of Texas water, river authorities are underscored exclusively here. Table 4.1 shows the various river authorities' wholesale water rates in 2004 for municipal, industrial, and agricultural uses.

This table overviews disparities but does not reveal fully the anomalistic character of these rates or the more careful reporting of the original table. For example, the following are three interesting, yet untabulated, details:

- Some of the reported agricultural rates pertain to interruptible water supply contracts that are less secure than water procured at higher rates. "Interruptible" water is agreed to be unavailable from the authority during periods of drought.
- The LCRA charged a "reserve rate" of $52.50 for water earmarked for its clients but not consumed, thus providing growing communities with a secure future water supply and the LCRA with revenue for funding water supply undertakings (recall the LCRA's recent $75 million purchase).
- The rate reported for the Trinity River Authority is applicable only to a single project supply area, and that for the Upper Colorado River Authority to a single city client.

Of the 15 river authorities listed in this table, 4 do not supply water and 1 does not supply wholesale water. The remaining 10 supply differing sectors. Two have

Table 4.1 *Wholesale water rates in 2004 (per acre-foot)*

River authority	Municipal	Industrial	Agricultural
Sulphur		not engaged in water supply	
Sabine			
Gulf Coast Division	$96.77–$30.64	$106.50–$38.06	
Toledo Bend Division	——————— $73.70–$25.48 ———————		
Red		not engaged in wholesale water supply	
Angelina Neches		not engaged in water supply	
Lower Neches	$51.35	$56.19	$15.00
Trinity	——————————— $75.00 ———————		
San Jacinto	——————— $73.70–$55.30 ———————		
Brazos	——————— $45.75 ———————		
Lower Colorado	———————————— $105.00 ———————		
Lavaca Navidad	——————— $93.87 ———————		
Guadalupe-Blanco	———————————— $88.00 ———————		
Upper Guadalupe		not engaged in water supply	
Upper Colorado	$140.00		
San Antonio	$84.00		
Nueces		not engaged in water supply	

Note: To convert $ per acre-foot to $ per 1,000 gallons, multiply the given price by 0.003; to obtain $ per 1,000 cubic meters, multiply by 0.811. For example, $100 per acre-foot is equivalent to $0.30 per 1,000 gallons or $81.10 per 1,000 cubic meters.

Source: Abridged from a more detailed table appearing in Sushma Krishnamurthi, Water Supply Aspects of River Authorities in Texas (master's thesis, Texas A&M University, 2006), 65–66.

ranges of rates given, as a consequence of decreasing block rates they apply to municipal and industrial uses. Although the ranges of these block rates appear to be wide, the highest prices of these schedules decline quickly as consumption rises. That is, most of the rates in these blocks are closer to the lowest price listed.

To generate some insights regarding the sensitivity of these rates to climate, Table 4.1 lists river authorities in a southwesterly order, beginning at the eastern Texas border. Thus, average precipitation is declining and temperature is rising as one proceeds down the table. At a rough level of approximation, it appears that more arid areas do experience higher wholesale water rates. Unfortunately, the mechanism by which this signaling is accomplished is not as complete as it could be, because river authorities do not incorporate the full value of natural water in their rates. This matter is revisited later in the chapter, for the issue also applies to water retailers.

Greater formality exists between water wholesalers and their clients than between retailers and their customers. River authority wholesalers establish written contracts with their clients, with each authority customarily applying its own standardized and evolving format. These contracts can contain an array of terms, including nonprice obligations such as environmental duties, planning expectations, and dispute resolution procedures. Rates and rate-updating procedures are incorporated as well. Some of these contracts are lengthy documents.[33]

Although the practice conflicts with efficiency ideals, publicly managed water rates in parts of the world are supportive of higher commercial and industrial water rates than household and agricultural rates. The reasoning behind these differential rates in some countries is welfare based and commonly results in the subsidization of particular water uses at below-cost rates. The data in Table 4.1 suggest that wholesale rates are more intersectorally level in Texas. Moreover, even when rates are unequal, as in the case of lower irrigation prices, the security of that water during drought is often lower than that of other sectors—justifying, to some extent, cheaper irrigation water.

Retail Water Pricing

As a result of differences in natural water's cost, distance from demand centers, and quality, retail water prices vary widely within Texas. Differences also are injected by variable accounting procedures, local policy, and cross-subsidization. Water suppliers are not required to report their rates to a Texas agency, so it is difficult to assemble a complete picture. Rates are changing regularly too. Many urban utilities may update their rates annually. Not only are rates subject to change, but structural elements change as well. For example, decreasing block rates have fallen out of favor in Texas over the past 25 years, and most retail suppliers now use a uniform rate or increasing block rates.[34] Cities tend to be interested in what rates other cities use, so the Texas Municipal League (a professional association) regularly tabulates water bills for different consumption levels in major cities.[35] Because actual bills have two parts—a flat (consumption invariant) fee and a volumetric (metered) water use charge—it is difficult to gain a full sense of the underlying rates from such compilations. As reported above for wholesaling, Texas retail prices do not commonly vary by sector except that larger meters (i.e., larger pipes) are typically charged higher flat fees, some manufacturers may be able to receive less-processed water at a lower rate, and block rate structures may affect sectoral prices in that consumption quantities may differ across sectors.

A recent comprehensive study of urban water demand in Texas yields more specific information about residentially oriented pricing. Figure 4.3 reproduces a graphic from this study pertaining to two classes of water suppliers. One type provides customers with both water and wastewater collection services, as typifies an urban area. The other provides only retail water, as typifies lower-density areas where septic systems are the prevalent means of wastewater disposal. Data for this graph comes from five years of monthly water use and water rates for several hundred Texas retailers. Using this information, bills can be calculated for the locale's average household.[36] Averaging across all 60 periods and retailers yields the information displayed in the figure's paired pie charts.

Note that both types of suppliers receive a sizable portion of their revenues from flat monthly fees, but wastewater service providers are less dependent on these fixed charges (36.3% versus 49.1%). This is an important distinction, because it affects scarcity signaling. Whereas suppliers prefer the revenue dependability of flat fees, the message that consumption-invariant charges send to consumers is that water has a zero price when they are contemplating water use changes and conservation investments. It is the volumetric portion of bills that suggests to consumers how vigilant they should be in their conservation behaviors.[37] If we

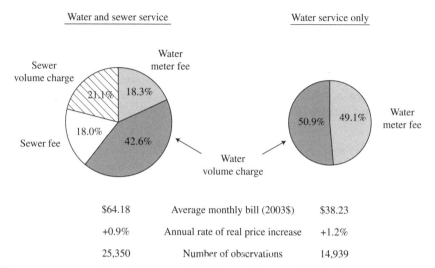

Figure 4.3 *Texas household water bills by service type, 1999–2003*

Source: David R. Bell and Ronald C. Griffin, *Community Water Demand in Texas as a Century Is Turned* (College Station, TX: Texas A&M University, 2006), 25.

look exclusively at the marginal price of water in this dataset and focus that inspection on the level of water use by the entity's average consumer, the price extends from $0 to $17.55 per thousand gallons.[38] This is a wide range that is likely to be more reflective of policy variability than scarcity variability across the state.

For example, if permanent water right values for natural water range from $0 to $10,000 per acre-foot across the state, then that may suggest a difference of $1.20 per thousand gallons creditable to unprocessed natural water.[39] No surface-water markets in the state have produced values as great as $10,000, so $1.20 is a confident upper bound. On the other hand, consideration of depletable (not Edwards) groundwater values, yet to be reflected by efficient markets, could be higher than current surface-water rates. Unfortunately, a complete appraisal of the role of natural water value in retail pricing has yet to be conducted for the state.

A PRICING ASSESSMENT

As in the case of water markets, no single, correct retail water price exists within Texas. Retail water price should vary in response not only to the variable scarcity of natural water, but also to the variable costs of transforming natural water into what is received by clientele. This implies the need for variability in retail prices across both time and location. Temporal variability of prices remains simplistic in Texas, as it is in other parts of the United States. Some retail suppliers maintain separate winter and summer rate schedules, which appear to be partially respectful of heightened summer scarcity. More refined (e.g., monthly) rates are not used in the state.

From a procedural perspective, rate design in Texas is very responsive to transformation costs, and it is fairly ignorant of water scarcity costs. Only in two cases—groundwater contracts and market acquisitions of additional water supplies—might retail prices begin to include raw water values as required for motivating efficient water use. Even these situations are inadequately addressed by contemporary ratemaking practices.

When a Texas water supplier incurs costs for water supply expansions, such as water right purchases in the Edwards or lower Rio Grande regions, these costs are generally able to affect the supplier's rates. That is, because these costs add to operation costs, or bond funds are used to defray these purchases, some portion of rate revenue must be made available to cover these costs. The same is true of groundwater contracts, because water suppliers are paying landowners for aquifer access rights. These costs then become part of the cost of service on which rates are founded. Unfortunately, these procedures are only partially successful in causing raw water values to be reflected in water-pricing signals. Some of the procedural failures are as follows, with the general result that water tends to be overused and misallocated in the state:

- When a water supplier incurs no costs for unimproved natural water, retail rates disregard natural water value completely. In these cases, rates are composed entirely of value-adding costs even where water scarcity exists.
- Even when bond indebtedness is incurred to pay for a purchased surface-water right, the cost is fundamentally a onetime cost. Rates are affected, but the impact disappears once the purchase price is reimbursed by ratepayers.
- Water suppliers carry substantial inventories of water right assets, but these assets are commonly unvalued, having no impacts on water rate signals, except for the short-term impact occurring when a new water right is purchased.
- A permanent surface-water right is a nondepreciating asset, because it is physically renewed by the hydrologic cycle every year. By retaining such rights, a utility is experiencing a lost revenue opportunity. Although recognition of this fact ideally would result in price signals that include natural water right values, water providers in Texas (and in other states too) do not commonly acknowledge these asset values in ratemaking. A few Texas urban utilities may have transitioned out of this problem by including all of their water right values in cost-of-service calculations.
- When a water retailer obtains all or part of its water inputs from a wholesaler, the completeness of water rate signals depends on whether the wholesale price successfully incorporates natural water values. Unfortunately, most wholesale water is priced on the basis of the accounting costs actually incurred by the wholesaler. In most instances, natural water costs are neglected here too. Even when a wholesaler incurs costs for natural water, such as in the LCRA's purchase of the Garwood system, the cost impact may be fleeting in that it disappears as the purchase price becomes fully reimbursed.

- Groundwater contracts are long-term instruments with use-based costs for the buyer, implying that groundwater costs are being continually passed on to final consumers. This would be a strong pricing accomplishment if contract costs fully reflected the social value of groundwater. Yet Texas sellers of groundwater access do not require compensation for all social costs. An important omission noted previously is the impact on hydrologically affected pumpers of the same aquifer, and an understated cost is the value of depletable groundwater in the future (see note 22).

Overall, then, the existence of Texas water markets has a positive influence on water pricing, insofar as markets affect costs, which influence rates. Yet disconnects exist where marketing does not occur or does not achieve correct values, or market prices are ignored partially or wholly. Such signaling failures are not unique to Texas, but they are problems nonetheless.

ADMINISTRATIVE OVERSIGHT OF MARKETING AND PRICING

The trading and pricing freedoms currently available to Texas water right holders and water suppliers are clearly the result of past legislative, judicial, and administrative decisions. Certain types of water rights are limited quantitatively by the state while others are not, and the state long has sanctioned trade in these rights. With respect to pricing decisions, water supply organizations have rather broad latitudes in setting rates, although certain principles have been loosely established by the state. The following sections identify major, ongoing features whereby Texas manages these marketing and pricing practices at the state level.

Marketing Oversight

Good water markets begin with quantitatively specified and enforced property rights to divert water from natural watercourses. Thus, a primary oversight role of the state is continued monitoring and enforcement of existing water rights. In addition, some of the quantitative restrictions expressed in these property rights may be revealed to be deficient in the future, thereby requiring modification through further state-level policy deliberations. Examples considered in other chapters include rising attention to freshwater estuary inflows across the Texas coast and the strong possibility that Edwards-fed springs are inadequately protected by the 572,000 acre-foot limitation on aggregate pumping.

Another important category of market oversight is the normal process whereby individual, trader-proposed exchanges gain state authorization. Water is not like an ordinary market good. Because of the hydrologic interrelatedness of water users in each watershed, there is justification for an administrative process whereby traders obtain state consent prior to consummating their deals. The important matter for the state to figure out is the seriousness of impacts on people other than the immediate traders.

The Texas process for approving transfers of appropriative surface-water rights is briefly summarized by the following points:

- Because water market exchanges commonly involve a change in the place or type of use, traders make formal application to "amend" the water right they wish to transfer.[40]
- Once an amendment application is ready to be considered by the administering agency (TCEQ), the agency mails a notice to other potentially concerned parties, including other right holders in the basin or basins of interest and "other persons who ... might be affected."[41] The purpose is to alert potentially affected water users and give them an opportunity to lodge objections to the transfer.
- If the amendment proposes an interbasin transfer of water, then application requirements are considerably more demanding[42] and a wider number of individuals are notified via mail and published announcements.[43] In this case, notification is extended to elected officials for counties and cities in the basin of origin as well as other leaders and representatives.
- The agency holds hearings regarding the proposed transfer and issues a ruling pertaining to its approval. The transfer might be rejected or fully approved, or it may be partially approved or have specific stipulations made pertaining to use of the amended water right. For example, the permitted withdrawal amount may be modified or the withdrawal pumping rate conditioned on streamflow level.

Clearly, this is not a costless process, for either the state or the participants.

There are exceptions to these amendment rules, and other interesting facets of marketing oversight arise as well. For example, simplifying circumstances specific to the lower Rio Grande (see note 9) and the Edwards Aquifer limit the need for amendment oversight, thereby exempting these water rights from this amendment process and allowing more rapid authorization. Canal company purchases are yet another variation; these represent changes in ownership but do not involve immediate transfers in the sector or place of water use, so such procedures need not be undertaken until such time as the buyer wishes to transfer place or sector of use. Transfers of rule-of-capture groundwater do involve third-party impacts, but protections against such effects have only begun to be explored within the state. Hence land-based transfers such as groundwater ranching and contracting are not regularly administered, meaning that state authorization is not required. This is an area in which groundwater conservation districts, many of which are quite young, have initiated new procedures, particularly when trades or contracts propose to export water out of the district.

It is noteworthy that the TCEQ-conducted amendment procedure applies to appropriative surface-water rights in the state, which is an arena where marketing activity is very thin, as noted previously. Recently, the increasing requirements of these rules have raised the transaction costs faced by traders, making it more difficult for trading to happen and reducing the benefits of those few trades that do occur. For example, a new requirement is that amendment applications "must be accompanied by water conservation and drought contingency plans,"[44] which is a surprising burden to place on an instrument (marketing) that by its very nature seeks to guide water to more efficient uses. Trading is a most important form of conservation because of its social service, so needlessly raising the costs of trading actually curbs conservation efforts.

Pricing Oversight

Whereas the state does not set water or wastewater rates, it does prescribe guidance for how rates should be determined. To a large extent, this guidance is consistent with the ordinary practices employed in other U.S. states, as well as those recommended by the American Water Works Association in its periodically updated M1 manual.[45] The state also assumes an appellate role in periodic rate revisions initiated by suppliers. That is, except for privately owned (for-profit) utilities, which compose less than 10% of Texas water suppliers and an even smaller percentage of water use, the oversight agency can hold hearings pertaining only to rate *appeals*.[46] So the jurisdictions of the individual suppliers have primary responsibility for deliberating and resolving rates. Only appeals of those rates can be heard by the TCEQ. Jurisdiction and procedural details regarding rate oversight are accessible online,[47] as is the rest of the Texas Water Code.

Specific Texas rules pertaining to acceptable rate design practices are contained in the Texas Administrative Code,[48] which states that rates are to be based on cost of service, and that a normal rate of return on invested capital is a permissible element of rates. These rules are fairly detailed in that certain cost items are clearly excluded or admitted in terms of their relevance to cost of service or invested capital appraisals. However, no explicit attention is given to the potential inclusion of unpaid natural water values. Indeed, the focus of these rules on accounting, and not economic, concepts of costs and returns implies that natural water values do not have allowable effects on rates until these values are experienced as tangible costs by the supplier. Thus, until a secondary path, such as water marketing, brings natural water values into the cost-of-service or asset valuation accounts of suppliers, water rates will not begin to properly reflect embedded water values.

CONCLUSIONS

Recent Texas water-planning efforts have asked for $30 billion to fund new water management strategies, with additional water development projects responsible for most of the proposed costs.[49] It is crucial to compare the merits of these high-dollar requests to the promise of available signaling policies. Whenever water policy is successful in allowing people to experience the scarcity value of water, for either gain or loss, the alleged need for grandiose construction programs is substantially curtailed, and financial resources for funding truly beneficial projects are obtained from users, rather than from general tax revenues. For these reasons, Texas's five distinct water markets, the state's practices of wholesale and retail water pricing, and the linkages among these institutions yield crucial lessons as well as calls for specific policy improvements.

Even though the novelty of water markets has faded with their rising acceptance in other states and countries, the varying foundations and conditions of Texas water markets offer interesting perspectives. Some of these markets have been operating for many decades. It is generally found that water transactions in all five markets have acted to accommodate population growth, assist economic development, and reduce water development activities. The Edwards groundwater market is quite young, yet it too seems destined to help in these same ways. These are noteworthy achievements, undoubtedly yielding large benefits to the people of Texas. Continuation and expansion of this market activity should eliminate a significant portion of the new projects being requested by the state's water-planning agency.

Where hydrology greatly reduces the effects of trades on third parties, surface-water markets perform admirably, as in the lower Rio Grande. Yet the lack of surface-water market activity in the rest of the state suggests that the broadest water market in Texas may be providing poor public service. To some extent, surface-water markets outside the Rio Grande have been upstaged by the acquisition of canal companies by river authorities. Such bulk transfers interfere with normal market development, but they also establish an alternative path whereby irrigation water can later be transferred to growing communities, via delivery contracts. This alternative has not been demonstrated to be a perfect substitute for open market transactions, given the large costs and intrinsic politics of water reallocation for river authorities.

In some ways, state policy appears to be backsliding in its support of water markets. Transaction costs that must be paid by traders are being increased unjustifiably. For example, new requirements insisting that traders submit conservation plans impose wasteful costs and render some trades undesirable. Trading is the pursuit of conservation, so why burden traders with unnecessary costs? Perhaps policymakers lack understanding of the subtleties of market-stimulated behavior and the importance of accurate signaling.

So in spite of their general accomplishments, certain markets are under-performing. In other ways, some Texas water markets are completely disappointing. Rates of aquifer depletion are not being well managed by existing land-based groundwater markets (see Chapter 9 for an important example). Nor is this system respectful of its effects on other groundwater users. Although these problems have spurred the legislature to establish groundwater management districts, which are apt to initiate regulatory controls, other options involving property right creation remain available. Yet after one groundwater district selected this approach, it was legally challenged, and the resulting judicial opinion has nullified the idea for now.[50] Thus, as demonstrated with policy reform for the Edwards Aquifer, replacing the rule of capture with true private property in groundwater is difficult. Even in a pro-marketing state such as Texas, it can be hard to activate the power of markets to reallocate scarce water.

Another disappointment arises from the callous treatment water markets extend to public good uses of water. Instream flows and freshwater inflows to coastal waters are not being championed by traders, nor should this be expected, given that public good uses of water are indeed public. Hence among these Texas lessons are examples that water markets are not definitionally efficient, nor are they definitionally more efficient than other policy options. Part of the crafting of good markets involves how these markets are to interface with other water policies and how social values beyond the concerns of traders are to be advanced or at least protected.

On the pricing front, Texas policies are not supportive of good signaling, though it can be claimed that the state's water markets have rendered the signals "less bad" than they are in other places. Economic doctrine informs us that water prices at all wholesale and retail levels should include the value of natural water, just as gasoline prices should include the value of crude oil. Otherwise, consumers of all types are not going to adopt efficient conservation practices, nor are they going to use efficient amounts of water.

While it is found that wholesale and retail prices are geographically variable in the state, evidence does not suggest that rates include the natural water values. Nor does the state have guiding procedures in place to encourage its cities, districts, authorities, and other water suppliers to set appropriate prices. So it comes at no surprise that water prices are generally understated throughout Texas. Implications for total water withdrawals and pumping are qualitatively obvious, even though the exact degree of underpricing is uncertain at this time. Also obvious are the impacts on the amount of water left instream for public good uses and the amount of groundwater conserved for future periods and future people. Thus, underpricing is sponsoring poor stewardship.

Although individual water suppliers may have the discretion to employ a more socially responsible model of water pricing, they are unlikely to do so until they are coerced by new state policy. When and if this occurs, water markets may produce yet another advantage, for water right prices may aptly reveal the value of natural water as it varies from place to place and time to time. This will be important information, especially where rates are set using procedures that acknowledge the underlying value of water.

NOTES

1. For example, city utilities that supply electricity as well as water are expected to apply revenue from electricity rates only to electricity provision. Also, general tax revenues derived from sales or property taxes are not generally applied to water supply costs.

2. Reviewed by Ronald C. Griffin, *Water Resource Economics: The Analysis of Scarcity, Policies, and Project* (Cambridge, MA: MIT Press, 2006), chs. 4, 8; and National Research Council, *Water Transfers in the West: Efficiency, Equity, and the Environment* (Washington, DC: National Academy Press, 1992).

3. It can be argued that the current system of reservoirs and control structures along a river modifies the flow regime so that no waters are fully natural any longer. However, the existing system cannot be bypassed or undone; it now defines the character of natural water.

4. In recent times, Texas's water rights administration agency has also issued term permits (generally expiring after 10 years, yet possibly renewable) for basins in which additional water appears to be available in the short run but may not be available in the long run as older rights

become more fully utilized. Even if such term permits had been granted permanent status, they would be very junior and unreliable. They are less valuable as a consequence.

5. Ronald C. Griffin and Gregory W. Characklis, Issues and Trends in Texas Water Marketing, *Water Resources Update* 121 (January 2002): 29–33.

6. Texas Administrative Code, Title 30, *Environmental Quality,* §§ 303.341–303.344, http://info.sos.state.tx.us/pls/pub/readtac$ext.viewtac (accessed June 2, 2010).

7. The region's recent water-planning studies indicate that 85% of water rights in the lower four counties are for irrigation (NRS Consulting Engineers, *Final Plan for Review: Rio Grande Regional Water Plan 2006,* 3–37, available at http://www.twdb.state.tx.us/rwpg/main-docs/2006RWPindex.asp, accessed June 2, 2010). However, this percentage is overstated when it is recognized that the quantity of irrigation rights is theoretical insofar as irrigation rights are incompletely fulfilled during dry years. On the other hand, during very wet years, "no-charge" water is declared to be available, and irrigators are the primary users of this water. "No-charge" means that any use is not recorded as debits (subtractions) from users' water right accounts.

8. It is still true, however, that upstream salinity sources have motivated urban entities to entertain expensive desalination plants.

9. The absence of return flow impacts of trade on third parties implies that traders' *personal* interests are more likely to capture all the important *social* interests associated with a given trade. Thus, privately arranged trades are more likely to improve public welfare in this case. The absence of monopolistic power influences means that individual entities do not have ability to dictate price, a technical condition required for the economic efficiency of marketplaces.

10. Texas Senate Bill 3, § 49.5, 80th Texas Leg., Regular Session, 2007, available at http://www.legis.state.tx.us/tlodocs/80R/billtext/html/SB00003F.htm (accessed June 2, 2010).

11. John A. Adams, Jr, *Damming the Colorado* (College Station, TX: Texas A&M University Press, 1990); James H. Banks and John E. Babcock, *Corralling the Colorado* (Austin, TX: Eakin Press, 1988); Floyd Durham, *The Trinity River Paradox* (Wichita Falls, TX: Nortex Press, 1976); Kenneth E. Hendrickson, Jr, *The Waters of the Brazos* (Waco, TX: Texian Press); U.S. Army Corps of Engineers (USACE), *Water Resources Development in Texas,* 1995.

12. Ronald C. Griffin, Gregory M. Perry, and Garry N. McCauley, *Water Use and Management in the Texas Rice Belt Region,* MP-1559 (College Station, TX: Texas Agricultural Experiment Station, Texas A&M University, 1984).

13. Jayson K. Harper and Ronald C. Griffin, *Regional Management of Water Resources: River Authorities in Texas,* MP-1666 (College Station, TX: Texas Agricultural Experiment Station, Texas A&M University, 1988).

14. Adams, *Damming the Colorado,* 107–08.

15. Texas Water Development Board, *A Texan's Guide to Water and Water Rights Marketing,* 2003, http://www.twdb.state.tx.us/publications/reports/WaterRightsMarketingBrochure.pdf (accessed June 2, 2010).

16. Ronald A. Kaiser, Texas Water Marketing in the Next Millennium: A Conceptual and Legal Analysis, *Texas Tech Law Review* 27 (1996): 181–261 (n. 10).

17. Griffin and Characklis, Issues and Trends.

18. Although Texas does not have a centralized conveyance system, smaller interbasin transfer systems owned by individual entities do exist, and more are under consideration. Texas Water Development Board (TWDB), *Water for Texas 2002* (Austin, TX: TWDB, 2002), figs. 5-18, 18-14, http://www.twdb.state.tx.us/publications/reports/State_Water_Plan/2002/FinalWaterPlan2002.asp (accessed June 2, 2010).

19. Charles E. Gilliland, Water Pressure: Below the Surface of GCDs, *Tierra Grande* 13 (April 2006), available at http://RECenter.tamu.edu/pdf/1770.pdf (accessed June 2, 2010).

20. As the Edwards Aquifer scenario in the next section illustrates, social choice regarding groundwater law is not limited to policies that can coexist with the rule of capture. It is also possible to reject the rule entirely and replace it with a true private property law. For aquifers with significant unrenewed storage volumes, the Edwards reform model is inadequate. Yet some literature indicates that it practical to assign rights to water in storage as well as regularly recharged water (see Griffin, *Water Resource Economics*, sec. 4.9). Such approaches may be very advantageous relative to the regulatory leanings of groundwater districts. Clay J. Landry, *A Free Market Solution to Groundwater Allocation in Texas* (Bozeman, MT: Political Economy Research Center, 2), available at http://www.texaspolicy.com/pdf/2000-12-01-environ-water.pdf (accessed June 2, 2010).

21. See, e.g. the Mesa Water Inc. website, http://www.mesawater.com. These arrangements promise to be subsidized by Texas taxpayers through tax exemptions to groundwater suppliers, meaning that nonbeneficiaries are bearing some of the costs and full-cost pricing signals are not being received by water users.

22. Technically, in depletion settings, a social value known as "marginal user costs" ideally would be factored into water deals (Griffin, *Water Resource Economics*, secs. 3.16, 4.9). However, the common property "feature" of rule-of-capture groundwater implies that the landowner does not have a secure title to future groundwater. So if a landowner makes the personal sacrifice necessary to conserve water for future use, there is no guarantee that the conserved water will still be there in the future. Hence landowners are likely to behave as if marginal user costs are near zero, with negative implications for both their water use decisions and their contracting of groundwater access. Consequently, a bias exists toward selling groundwater access too cheaply, resulting in overuse and too-rapid depletion.

23. Edwards Aquifer Authority, *Comprehensive Water Management Plan 2004*, 36.

24. The first phase pumping limit of 450,000 acre-feet expired at the end of 2007, after which the stricter limit was to be in force. The maximum allowed pumping level may be increased if the oversight agency finds, "in consultation with appropriate state and federal agencies," that a higher level may be available. Texas Senate Bill 1477, § 1.14(d), 73rd Texas Leg., Regular Session, 1993, http://www.legis.state.tx.us/tlodocs/73R/billtext/html/SB01477F.htm (accessed June 2, 2010).

25. Edwards Aquifer Authority, *Comprehensive Water Management Plan 2004*, 99.

26. Robert A. Collinge, Peter M. Emerson, Ronald C. Griffin, Bruce A. McCarl, and John D. Merrifield, *The Edwards Aquifer: An Economic Perspective*, TR-159 (College Station, TX: Texas Water Resources Institute, 1993), available at http://ron-griffin.tamu.edu/reprints/Collinge Etal1993.pdf (accessed June 2, 2010).

27. The 1993 statute directs the authority to "distinguish between discretionary and nondiscretionary use" in administering water use during "critical periods," thereby erecting a rough hierarchy.

28. Gregory M. Ellis, Edwards Aquifer Authority (presented at Straddling the Divide conference, Chicago, February 15–16, 2005), available at http://www.nipc.org/environment/slmrwsc/conferences/5A_Ellis.pdf (accessed June 2, 2010).

29. In large part, this program was intended as an insurance measure following a 1996 drought, but because of wet weather conditions in 1997, the suspension program had relatively little effect on aquifer levels or springflows. Only an estimated 3,868 acre-feet of water was conserved by the program, 25% of what would have been conserved if 1997 had turned out to be an average year. Keith O. Keplinger, Bruce A. McCarl, Chi Chung Chen, and Ruby Ward, *The 1997 Irrigation Suspension Program for the Edwards Aquifer: Evaluation and Alternatives*, TR-178: 12 (College Station, TX: Texas Water Resources Institute, Texas A&M University, 1998).

30. TWDB, Texas Water Trust, http://www.twdb.state.tx.us/assistance/WaterBank/wtrust.asp (accessed October 7, 2006).

31. In some ways, this is a challenging objective to achieve. Many water suppliers also are engaged in the provision of other services for which fees are assessed, such as electricity production (LCRA) and distribution (many cities) and garbage collection. Proration of costs that commonly support multiple functions can be arbitrary in these contexts. As examples, how should dam maintenance costs be split between water supply and hydroelectric activities, and how should the administrative salaries of a city utility be prorated?

32. Griffin, *Water Resource Economics*, ch. 8.

33. See the interesting and well-detailed framework for Lower Colorado River Authority contracts at LCRA, "Firm" Water Supply Contracts, http://www.lcra.org/water/supply/contracts/index.html (accessed June 2, 2010).

34. Texas policy now discourages the use of decreasing block rates. For example, it is suggested that municipal participants in certain state programs reject "flat rate or decreasing block rates" in their required water conservation plans. Texas Administrative Code § 288.2(a)(3).

35. See Texas Municipal League, TML Surveys, http://www.tml.org/surveys.asp (accessed June 2, 2010).

36. Each observation, therefore, applies to a single month and a single supplier. There can be as many as 60 observations per supplier, and an inflation adjustment was incorporated in the calculations.

37. For such incentives to work well, it is helpful if suppliers are clear about billing procedures in the monthly bills they send to their consumers. All aspects of applicable rates, including what is fixed and what is volumetric, should be readily apparent. This goal has yet to be achieved in much of the state, but considerable progress has been made, particularly in more progressive towns and in areas where rising scarcity has underscored this need.

38. David R. Bell and Ronald C. Griffin, *Community Water Demand in Texas as a Century Is Turned* (College Station, TX: Texas A&M University, 2006), 27, available at http://ron-griffin.tamu.edu/reprints/udemand2006.pdf (accessed June 2, 2010).

39. At a 4% rate of discount, the permanent value of water is about 25 times greater than the annual (i.e., lease) value of water (Griffin, *Water Resource Economics*, 210).

40. Texas Administrative Code, § 295.9(4).

41. Ibid., § 295.153(b).

42. Ibid., § 295.13.

43. Ibid., § 295.155.

44. Ibid., § 295.9(4).

45. American Water Works Association (AWWA), *Principles of Water Rates, Fees, and Charges*, 5th ed. (Denver: AWWA, 2000).

46. Texas Commission on Environmental Quality, *TCEQ Jurisdiction over Utility Rates and Service Policies*, RG-245, http://www.tceq.state.tx.us/comm_exec/forms_pubs/pubs/rg/rg-245.html (accessed June 2, 2010).

47. Texas Water Code, Title 2, *Water Administration*, ch. 13 (August), http://www.statutes.legis.state.tx.us/Docs/WA/htm/WA.13.htm (accessed June 2, 2010).

48. Texas Administrative Code, esp. §§ 291.31–291.32.

49. TWDB, *Water for Texas 2007 (Draft)*, vol. I (Austin, TX: TWDB, 2006), 6 http://www.twdb.state.tx.us/publications/reports/State_Water_Plan/2007/Draft_2007SWP.htm (accessed June 2, 2010).

50. Supreme Court of Texas, *Guitar Holding Company v. Hudspeth County Underground Water Conservation District No. 1, et al.*, 2008, http://www.supreme.courts.state.tx.us/historical/2008/may/060904.pdf (accessed June 2, 2010).

CHAPTER 5

The Edwards Aquifer: Hydrology, Ecology, History, and Law

Todd Haydn Votteler

T he Edwards Aquifer is essentially the sole source of water for almost 2 million people, including the residents of the city of San Antonio and the surrounding region. Currently, 95% of San Antonio's water use is from this aquifer.[1] Management of the Edwards Aquifer has been a controversial and divisive issue for more than 50 years. Recent changes have effectively transformed aquifer management from the common property style to private property, so some measure of protection is now extended to springflows. Moreover, market-based exchanges now carry out the growth-driven reapportionment of water to the urban sector. The trials and accomplishments of this story isolate a unique model, and perhaps these lessons will assist other regions confronted with groundwater overuse.

Because of the aquifer's substantial contribution to the flow of regional rivers through springs and the unique forms of endemic life in its springs, its use as a water source has been the focus of intense regional competition. Conflicts have erupted between rural and urban interests and between those dependent on aquifer pumping and those living downstream of its spring outlets who depend on springflows for their surface water. These conflicts have played out in local, state, and federal courts, as well as the Texas legislature, resulting in changes that have radiated out across the state through Texas water law. To resolve the basic issue of aquifer water apportionment, some have demanded regulation of Edwards Aquifer groundwater withdrawals, while others have contended that such limitations would violate private property rights under the Texas Constitution[2] and the Fifth Amendment to the U.S. Constitution.[3] The Endangered Species Act (ESA) eventually brought state regulation to the Edwards Aquifer and an end to unrestricted withdrawals of groundwater from the aquifer.[4]

HISTORY

Humans have relied on Edwards Aquifer springs for more than 12,000 years.[5] The springs were an important resource for the earliest inhabitants of the region. San Antonio, New Braunfels, San Marcos, and Uvalde were founded around Edwards Aquifer springs long before wells were drilled into the aquifer. The use of artesian wells from the aquifer dates back to 1884, when the first irrigation well was drilled in Bexar County.[6] Groundwater pumping began in earnest during the 1950s.[7] Today the Edwards Aquifer supplies high-quality water to municipal, agricultural, industrial, and recreational users.

The quality and quantity of water supplied throughout the historical period of use have been so high that San Antonio relied on the aquifer as its only source of water. The Edwards Aquifer was recognized by the U.S. Environmental Protection Agency (EPA) in 1975 as the nation's first "sole source aquifer" under the Safe Drinking Water Act of 1974.[8] Until the drought of record (1947–1957), the aquifer was so prolific, and consumption so minimal, that pumping from wells appears to have made little difference in spring discharge. Today, however, many of the springs, such as San Antonio Springs, rarely flow unless a flood event fills the aquifer.

HYDROLOGY

The Edwards (Balcones Fault Zone) Aquifer is divided into the northern, Barton Springs, and San Antonio segments. The San Antonio segment (Figure 5.1) is the subject of this chapter and what is referred to hereinafter as the Edwards Aquifer. It stretches about 200 miles, from Brackettville east to San Antonio and northeast to Kyle, where it generally has been believed that a groundwater divide separates the San Antonio and Barton Springs segments. However, recent research indicates that aquifer water flows past San Marcos Springs and then on to Barton Springs during drought conditions, suggesting that the management of the San Antonio segment influences aquifer levels for the Barton Springs segment and the discharge of Barton Springs.[9]

The San Antonio segment is one of the most permeable and productive karst aquifers in the United States.[10] Karst aquifers are limestone aquifers characterized by sinkholes, caves, and underground drainage systems. The Edwards Aquifer is very transmissive because of its highly permeable and porous limestone. In total, the aquifer encompasses a contributing zone of some 4,400 square miles, a recharge zone of 1,500 square miles, and a confined zone of 2,100 square miles, totaling about 8,000 square miles.[11]

An overly simple analogy to the complex Edwards Aquifer likens it to a bucket with different-size holes that represent the springs at several levels from top to bottom. If the bucket is full of water, the water flows out from all the holes at variable velocities, depending on the water level in the bucket and the size and elevation of the holes. As the water level declines, flow from each hole decreases until the lower edge of each downward hole is reached, when flow ceases. San Antonio, Comal, and San Marcos Springs are examples of holes in the bucket

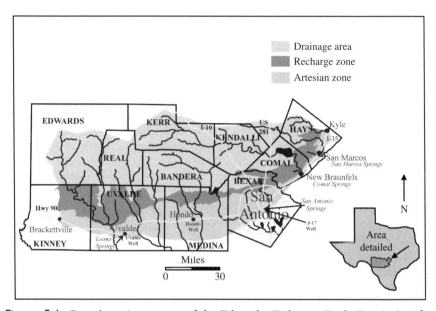

Figure 5.1 *San Antonio segment of the Edwards (Balcones Fault Zone) Aquifer*

Source: Adapted from Gregg A. Eckhardt, The Edwards Aquifer Website, http://www.edwardsaquifer.net (accessed June 4, 2010).

and are also the sources of rivers of the same names, all of which eventually flow into, and provide much of the base flow for, the Guadalupe River.[12]

The Springs

Comal and San Marcos Springs are among the largest springs in the United States. Comal Springs actually consists of 18 or more spring openings. San Marcos Springs consists of some 200 outlets that originate from three large fissures and many small openings at the bottom of Spring Lake.[13] Table 5.1 summarizes several important hydrologic characteristics of the Edwards Aquifer and its springs, as well as regulatory and management limits.

Withdrawals from the Edwards have increased from approximately 100,000 acre-feet (ac-ft) in 1934 to a peak of 542,400 ac-ft during the drought year of 1989.[14] As withdrawals from the aquifer multiplied, the possibility became greater that Comal and San Marcos Springs could become intermittent or even cease to flow altogether. Combined, these two spring systems annually contribute an average of about 333,800 ac-ft of water into the Guadalupe River. This annual springflow contribution to the Guadalupe is approximately twice the amount that the city of San Antonio currently uses from the aquifer. The largest supplier of water in San Antonio is the San Antonio Water System (SAWS). The average SAWS customer in the late 1980s was using 285 gallons per capita per day (gpcd) of water.[15] However, by 2003, the Texas Water Development Board estimates that water use for the city had dropped to 142 gpcd, which is among the most efficient

Table 5.1 *Key Edwards Aquifer pumping and recharge numbers*

Average Edwards Aquifer annual recharge (1934–2007)	732,400 ac-ft/yr
Median Edwards Aquifer annual recharge (1934–2007)	585,700 ac-ft/yr
Lowest recorded Edwards Aquifer recharge (1956)	43,700 ac-ft
Highest recorded Edwards Aquifer recharge (1992)	2,486,000 ac-ft
Average annual contribution of Comal and San Marcos Springs to flow in the Guadalupe River (1956–2007)	333,800 ac-ft/yr
State recommended limit on aquifer pumping in *1968 Texas Water Plan*	400,000 ac-ft/yr
Average annual aquifer pumping (1997–2007)	386,900 ac-ft
Median annual aquifer pumping (1997–2007)	379,900 ac-ft
Groundwater rights in original permit applications submitted to EAA	846,180 ac-ft/yr
Aquifer pumping allowed by Senate Bill 1477 before 2007 amendments by legislature	400,000 ac-ft/yr
Aquifer pumping allowed after 2012[a]	Not yet determined
Pumpable from aquifer during a repeat of drought of record and ensuring 150 cfs at Comal Springs[b]	165,000 ac-ft/yr
Pumpable from aquifer during a repeat of drought of record and ensuring 60 cfs at Comal Springs[b]	225,000 ac-ft/yr
Amount pumped in 1956 when Comal Springs ceased to flow for 144 days during the drought of record	321,000 ac-ft
Groundwater permits issued by EAA in 2007 (agricultural, 253,400 ac-ft/yr; municipal, 253,400 ac-ft/yr; industrial, 42,100 ac-ft/yr)	549,000 ac-ft/yr
Pumping authorized by legislature in 2007	572,000 ac-ft/yr

Notes: ac-ft = acre-feet; ac-ft/yr = acre-feet per year; cfs = cubic feet per second; EAA = Edwards Aquifer Authority
[a] This will be the amount ensuring continuous minimal flow of Comal and San Marcos Springs for endangered species according to the U.S. Fish and Wildlife Service (FWS)
[b] Texas Water Development Board (TWDB) model and FWS
Sources: U.S. Geological Survey (USGS), Estimated Recharge to the Edwards Aquifer in the San Antonio Area, Texas, 1934 to 2006 (2007); USGS, *Recharge to and Discharge from the Edwards Aquifer in the San Antonio Area, Texas, 1934–2007* (San Antonio, TX: USGS, 2008); Todd H. Votteler, Water from a Stone: The Limits of the Sustainable Development of the Texas Edwards Aquifer (PhD diss., Southwest Texas State University, 2000, on file with author); Edwards Aquifer Authority (EAA), Fact Sheet: Final Groundwater Withdrawal Permit Amounts Established (November 9, 2005); EAA, Edwards Aquifer Authority Hydrologic Data Report for 2006, Report No. 07-01 (July 2007); Geary M. Schindel et al., *Edwards Aquifer Authority Hydrologic Data Report for 2007*, EAA, Report No. 08-02 (August 2008), 19, 30; and Sam Vaugh, HDR Inc., personal communication, August 29, 2007.

in the state.[16] As part of the gains in efficient water use, water that is lost in transmission (unaccounted-for water) has decreased substantially, from about 22% of the Edwards Aquifer water pumped by SAWS in 1982 to just over 9% in 2007.[17] SAWS has set a goal of reducing use to 116 gpcd by 2018.[18] Until the Endangered Species Act (ESA) litigation resulted in limitations on aquifer pumping, little incentive existed for pumpers to spend money to plug leaks or individuals to conserve water. It was simply cheaper to pump greater amounts of water from the aquifer to overcome transmission losses, and to many, there appeared to be no real limit to the amount of water that could be withdrawn to satisfy growing demands in the region.

During droughts, the discharge from Comal and San Marcos Springs diminishes in total volume but increases in terms of its percentage of contribution to instream flows in the Guadalupe River and to freshwater inflows to the river's estuary and San Antonio Bay. As Figure 5.2 illustrates, the springs regularly provide the majority of flow in the Guadalupe River, as well as a substantial amount of the freshwater inflows to San Antonio Bay, during the frequent droughts that occur in the region.[19]

Water from the aquifer also supports the economies of agriculture-based counties west of the city, Comal and Hays Counties to the east, and counties in the Guadalupe River basin all the way to the Texas Gulf Coast. Permits issued by the state to surface-water rights holders in the Guadalupe River basin are based in part on discharges from the aquifer's springs. A majority of the surface-water right permits were issued before 1956, the year when withdrawals from the aquifer first exceeded 300,000 ac-ft/yr.[20]

The total volume of water circulating within the Edwards Aquifer is not known with certainty but has been estimated at 45 million ac-ft. However, much of this water is at depths that make its use currently uneconomical. Aquifer levels are dependent on highly variable annual rainfall, recharge (Table 5.2), and the rate of groundwater withdrawals (Table 5.1).[21] Much of the aquifer recharge occurs as the result of brief but intense storms west of San Antonio. This recharge occurs where three major rivers, the Nueces, the San Antonio, and Guadalupe, cross the aquifer recharge zone.

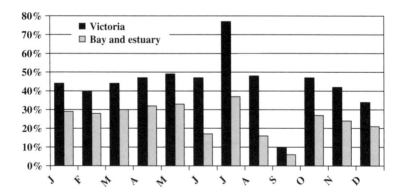

Figure 5.2 *1996 drought contribution of Comal and San Marcos Springs to Guadalupe River at Victoria and to San Antonio Bay and Guadalupe Estuary inflows*

Note: Channel losses were applied to U.S. Geological Survey (USGS) spring discharge data based on values from the Texas Commission on Environmental Quality's Water Availability Model.

Table 5.2 *Estimated Edwards recharge by decade*

Decade	Millions of ac-ft
1940–1949	4.7
1950–1959	4.7
1960–1969	5.6
1970–1979	8.9
1980–1989	7.6
1990–1999	9.7
2000–2009	9.7

Source: Calculations based on USGS, *Recharge to and Discharge from the Edwards Aquifer in the San Antonio Area, Texas, 1934–2007* (San Antonio, TX: USGS, 2010).

As the water flows eastward within the confined zone, wells intercept a significant portion of the aquifer's annual recharge.[22] The flow of water also is redirected through the Knippa Gap, northwest of Uvalde, an ill-defined geologic feature that restricts, to an unknown degree, the flow of water from the western parts of the aquifer to the east.

Scientists generally agree that somewhere south of the Edwards Aquifer downdip, a "bad-water line" separates the area of usable groundwater from an area where wells produce water of unacceptable quality. However, knowledgeable persons disagree as to the risk of this line moving as a result of withdrawing large quantities of water from the aquifer during dry years. The possibility of saline water encroachment has been a concern since the latter years of the drought of record in the 1950s, when residents reported that some freshwater wells on the southern edge of the aquifer experienced an intrusion of highly mineralized water. The bad-water line exists in close proximity to both Comal and San Marcos Springs where endangered aquatic species reside. The potential intrusion of saline water into the springs or groundwater wells during a prolonged drought could have dire consequences for the survival of spring-dependent biota and for groundwater users.

Drought and the Aquifer

In water supply planning, a most important question is how much water can be supplied during historical droughts. The challenge droughts represent to those who depend on the aquifer is made even greater in the absence of readily available water supply alternatives. Significant droughts and floods occur frequently in the Edwards Aquifer region.[23] Daily discharge of the Guadalupe River into San Antonio Bay has varied from flows approaching zero cubic feet per second (cfs) during the height of the drought of record in 1956 to 466,000 cfs during the peak of the flood of 1998. The resulting wide variations in recharge (see Table 5.1) present challenges to water supply planning in the Edwards region.

In Texas, the critical drought period used for planning and management purposes is called the drought of record, which is defined as the period of time when natural hydrologic conditions provided the least amount of water supply.[24] It is the standard for water resource planning in Texas,[25] as it is in most if not all western states. Most of the storage in Texas surface-water reservoirs is permitted on a firm yield basis, with the firm yield volume being the maximum quantity of water reliably available during a repeat of the drought of record.[26] For Texas and the Edwards Aquifer, the drought of record is the one that occurred from 1947 to 1957.[27] During this drought, Comal Springs ceased to flow for 144 days in 1956.[28] Although a total of 321,000 ac-ft of water was pumped from the aquifer that year, groundwater modeling indicates that in the absence of aquifer pumping, Comal Springs would have continued to discharge at about 300 cfs throughout the summer of 1956, providing insight on the impact of pumping from the Edwards Aquifer on Comal Springs discharge.[29] No physical or archeological record indicates that San Marcos Springs has ceased to flow during the last 10,000 years.

By the end of 1956, about 94% of Texas's 254 counties were classified as disaster areas for lack of precipitation. Recharge to the aquifer was below average for each of the 14 years from 1942 to 1956, with an average annual recharge of 300,600 ac-ft during this period compared with 732,400 ac-ft for the period of record, 1934 to 2007 (see Tables 5.1 and 5.2).[30] Estimates of how often a similar drought can be expected to occur range from once every 50 to 100 years.[31] Today the detrimental effects accompanying a drought of record would be far greater because of population and economic growth.

Climate Change and the Aquifer

Since the 1960s, the Edwards Aquifer region generally has been in a wet cycle.[32] Despite droughts in 1995–1996 and 1998, the 1990s had the highest total recharge of any decade during the period of record, greater than the total recharge for the 1940s and 1950s combined (see Table 5.2). This period of generally high recharge, during which withdrawals from the aquifer have reached their highest levels, eventually will be supplanted by an extended period of moderate to low recharge. In fact, the patterns of rainfall during the 1990s and 2000s have shown similarities in that record high, or near record high, recharge years preceded significantly dry years, providing high aquifer levels at the onset of drought. Also, each period was followed by a very high recharge year that allowed water levels in the aquifer to recover rapidly and rebound to above-average levels. Much of the population growth in the Edwards Aquifer region has occurred during the wet cycle that has characterized the last three decades, and the populace has been generally accustomed to a water surplus. During this same wet period, the Texas legislature has increased the maximum amount of water that can be pumped from the aquifer. The predicament this presents to the area is much like the issue the western states faced after Colorado River water was apportioned in 1922 based on a wet cycle that skewed estimates of available water upward to an amount greater than could be supported by the long-term average.

Although the potential impacts of climate change have yet to play a direct role in the management of the Edwards Aquifer, research by Loaiciga and colleagues has deemed the system "very vulnerable to climate-change impacts" based on a number of factors, including the region's near sole dependence on the aquifer to meet water demands; the scarcity of alternative water supplies; substantial climatic variability in the region; the increasing trend of groundwater use over the last 65 years resulting from economic and population growth; the presence of unique aquatic habitats supporting a variety of endangered species that face extinction under current trends of groundwater use; and the fact that local, state, and federal institutions are mired in a complex web of technical, scientific, and legal uncertainties for resolving aquifer management issues. This same study concludes that the Edwards Aquifer "is not a suitable sole-source water supply to meet the forecasted water demands" in the future, and that alternative water supplies should be developed and supplemented by water conservation and aquifer protection strategies.[33]

A 2001 study by Chen and others anticipates that the impact of climate change could be substantial, if not cataclysmic, projecting reductions in annual recharge to the Edwards Aquifer in drought, normal, and wet years ranging from 20% to 34% by 2030 and 32% to 49% by 2090.[34] Seager and colleagues report a broad consensus in climate models that the Southwest, including the aquifer region, will dry significantly in the twenty-first century, and that the transition to a more arid climate is already under way, with levels of aridity similar to those during the drought of record becoming the new climatology in the coming decades. Furthermore, the authors assert that the most severe future droughts will occur during persistent La Niña events and will be more severe than the drought of record or the historical Dust Bowl disasters.[35]

ECOLOGY

Water demands of the growing population of the Edwards region were once the sole determinant of the groundwater allocation, but today the unique aquifer-supported ecology is an important competing consideration.

Biological Diversity

The Edwards Aquifer is considered one of the most diverse aquifer ecosystems in the world.[36] Within the aquifer, species exist that are found nowhere else and about which little is known. For example, two species of unique blind catfish, called widemouth and toothless blindcats, are occasionally pumped out of the aquifer from depths as great as 2,135 feet.[37] The U.S. Fish and Wildlife Service (FWS) considers the Comal and San Marcos Springs ecosystems to contain one of the greatest known diversities of organisms of any aquatic ecosystem in the Southwest. This is in part because the constant temperature and flow of the aquifer's high-quality waters create unique ecosystems that support the development of species that are restricted geographically and do not occur elsewhere. Comal and San Marcos Springs constitute habitat for one threatened and seven endangered species listed by the FWS. The FWS recovery priority for these species indicates that each faces a high degree of threat and low potential for recovery, and that the survival of each species is in conflict with development projects or other forms of economic activity.[38] An additional 30 aquatic species associated with the Edwards Aquifer are considered candidates for listing by the FWS.[39]

When flows from the springs are reduced to critical levels, aquatic habitat is affected, causing "takes" of species listed under the ESA, and the flow of surface water downstream in the Guadalupe River decreases, reducing instream flows and freshwater inflows to San Antonio Bay. Extremely low flow or no flow from these springs places the continued existence of species in the spring ecosystems in "jeopardy." Under the ESA, the take of a threatened or endangered species by any person subject to the jurisdiction of the United States constitutes a violation of the act. "Take" means "to harass, harm, pursue, hunt, shoot, wound, kill, trap, capture, or collect, or to attempt to engage in any such conduct."[40] Withdrawals from the Edwards Aquifer for municipal, industrial, agricultural, recreational, and other uses contribute to the reduction of spring discharge at Comal and San Marcos Springs, which in turn can cause take and jeopardy of the listed species.[41] During the drought of record in 1956, the fountain darter was extirpated from the Comal Springs ecosystem when the springs ceased to flow for almost five months. Fountain darters were reintroduced into Comal Springs in the 1970s.

Ecological Pumping Constraints

During the *Sierra Club v. Babbitt* litigation over the protection of endangered species in the 1990s (discussed in more detail below), the FWS provided the U.S. District Court in Midland, Texas, with its "best professional judgment" of the flow and discharge rates at which the species of concern at Comal and San Marcos Springs are placed in jeopardy and at risk of take. As a result, presumed tripwires for an ESA enforcement action were set at flow rates of 200 cfs at Comal Springs, below which risk of take was determined to occur, and 100 cfs at San Marcos Springs, below which species are considered to be in jeopardy (take and jeopardy were set for Comal, but only jeopardy was established for San Marcos).[42] The endangered fountain darter at both Comal and San Marcos Springs is typically the first species to be affected by declining spring discharge, and therefore the darter population serves as an indicator of stress to the Edwards Aquifer system. The San Marcos and Comal Springs and Associated Aquatic Ecosystems (Revised) Recovery Plan acknowledges that the key issue to survival of the listed species is the conservation of the aquatic ecosystems at Comal and San Marcos Springs dependent on their flow, as well as the aquifer itself.[43]

LAW

The unregulated use of the Edwards Aquifer was replaced by a permit system and water market as the result of ESA litigation.

Groundwater vs. Surface Water Regulation in Texas

Historically, no limits were placed on groundwater withdrawals from the Edwards Aquifer.[44] Groundwater in Texas has been governed by the English common-law concept known as the rule of capture, the right of capture, or the law of absolute ownership, as well as other names.[45] In accordance with this rule, underground water can be withdrawn by an owner of the overlying land, even from beneath adjoining owners' land, unless a state statute specifies otherwise. In addition, an adjoining landowner whose available groundwater is adversely affected by someone else's pumping has no remedies in tort law unless waste occurs.[46] Under the rule of capture, gross misallocations of resources can occur. For example, in 1991, a catfish farm 15 miles southwest of San Antonio called Living Waters Artesian Springs Ltd began using a large amount of aquifer water—as much as 40 million gallons a day by some estimates—to raise catfish; the water was then discharged directly into the Medina River.[47] On an annual basis, this usage equaled approximately 25% of the city of San Antonio's total pumpage at that time.[48]

By contrast, surface water in Texas is governed by the appropriative water rights doctrine, also known as prior appropriation, which is common in most western states. Under this doctrine, surface water is held in trust by the state for the benefit of all the people, subject to a state-granted right to use. Those who are "first in time" are "first in right" to take or divert water from a surface watercourse or reservoir and apply it to a beneficial use.[49]

As coexisting legal frameworks, prior appropriation and the rule of capture encourage incompatible behaviors by water users, depending on the source. They contribute to the deleterious effects of droughts by treating surface water and groundwater as separate legal entities. The separation ignores the fundamental hydrologic connection between them and provides no incentives for their efficient conjunctive use. This legal and hydrologic dichotomy is a complicating factor for those with the responsibility for managing water in Texas—particularly the Edwards Aquifer and the Nueces, San Antonio, and Guadalupe Rivers, because of the degree of interaction among these systems. The legislature made an exception to the rule of capture in 1993 by creating the Edwards Aquifer Authority (EAA) to limit withdrawals to protect endangered species and guarantee minimum flows of groundwater from Comal and San Marcos Springs into the Guadalupe River.[50]

As surface water recharges the aquifer west of San Antonio, and then flows from Comal and San Marcos Springs into the Guadalupe River basin east of San Antonio, its legal character is transformed as it changes from surface water to groundwater to surface water again. Before the creation of the EAA, this transformation was from prior appropriation to rule of capture and then back to prior appropriation again. As previously noted, permits issued by the state to surface-water rights holders downstream on the San Marcos, Blanco, and Guadalupe Rivers are based in part on flows from the aquifer. Therefore, reduced recharge and increased pumping in the Edwards Aquifer region deplete the discharge of water at the springs, interfering with established surface-water rights of users in the downstream counties in the Guadalupe River basin.

Sierra Club v. Babbitt: Pumping Limits Mandated

The landmark legal case concerning the Edwards Aquifer has been *Sierra Club v. Babbitt*.[51] In 1991, the Sierra Club filed a suit in the U.S. District Court in Midland, Texas, alleging that Secretary of the Interior Bruce Babbitt and the FWS had allowed takings of endangered species by not ensuring a water level in the Edwards Aquifer adequate to sustain the flow of Comal and San Marcos Springs to protect the endangered species.[52] The Sierra Club was joined by the Guadalupe-Blanco River Authority, which had originally filed the notice of intent to sue under the ESA, and others. The plaintiffs requested that the FWS be enjoined to restrict withdrawals from the Edwards Aquifer under certain conditions and develop and implement recovery plans for named endangered and threatened species found in the aquifer and at Comal and San Marcos Springs.[53]

On February 1, 1993, Judge Lucius D. Bunton III ruled in favor of the plaintiffs and required the FWS to determine the minimum spring discharge requirements to avoid take and jeopardy of the listed species in both springs:

> I entered my judgment in January 1993 and essentially found that the overpumping from the Edwards Aquifer could indeed endanger the species that I had previously found were endangered in the Comal and San Marcos Springs. In the finding I expressly stated that the solution should be by the state rather than the federal government, and I would give the state an opportunity to address the matter in the coming session of the Texas Legislature.[54]

Bunton ruled that if the Texas legislature did not adopt a management plan to limit withdrawals from the aquifer by the end of its current session, the plaintiffs could return to the court and seek additional relief.[55] The Sierra Club indicated that if it had to return to the district court in 1993, it would seek regulation of the aquifer by having it placed under federal judicial control through the FWS.[56]

The Texas Legislature Creates the Edwards Aquifer Authority

Bunton issued a warning that "the next session of the Texas Legislature offers the last chance for adoption of an adequate state plan before the 'blunt axes' of Federal intervention have to be dropped."[57] Senate Bill (SB) 1477, or the Edwards Aquifer Authority Enabling Act, was adopted by the legislature on May 30, 1993, one day before the deadline for threatened federal action.[58] The act created a conservation and reclamation district, named the Edwards Aquifer Authority (EAA), replacing the Edwards Underground Water District, which was created by the Texas legislature in 1959 after the drought of record ended in 1957.[59] The EAA was charged with regulating groundwater withdrawals pursuant to the Conservation Amendment in the Texas Constitution, Article XVI, § 59, replacing the rule of capture in five counties and portions of three others with a permit system.[60] Under the act, annual withdrawals were to be limited to 450,000 ac-ft before December 31, 2007, and 400,000 ac-ft thereafter, unless drought conditions required more severe restrictions.[61] By December 31, 2012, "the authority [EAA] … shall … ensure that … the continuous minimum springflows of the Comal Springs and the San Marcos Springs are maintained to protect endangered and threatened species to the extent required by federal law."[62] The EAA is specifically charged by SB 1477 with protecting threatened and endangered species.[63]

Private Property Rights

Interests opposed to the end of unrestricted pumping from the Edwards Aquifer claim that their individual private property rights have become endangered. Some have contended that the regulation of Edwards Aquifer groundwater through the ESA is a taking of private property rights.[64] However, it is the regulation and allocation of Edwards Aquifer water that has actually created property rights. Until permits to withdraw specific amounts of water were issued by EAA, enforceable property rights, from a free-market perspective, did not exist in the Edwards Aquifer groundwater. This is because the fundamental characteristics of property rights were absent. In neoclassical economic theory, a "property right" refers to a bundle of entitlements defining the owner's rights, privileges, and limitations for use of a resource. These property rights can be vested with individuals, corporations, or the government. An efficient property rights system has the following characteristics:

- universality: all resources are privately owned and all entitlements completely specified;
- exclusivity: all benefits and costs accrued as a result of owning and using the resources should accrue to the owner, and only to the owner, either directly or indirectly by sale to others;
- transferability: all property rights should be transferable from one owner to another in a voluntary exchange; and
- enforceability: property rights should be secure from involuntary seizure or encroachment by others.[65]

In the Edwards Aquifer, none of these characteristics were present under the rule of capture. There was no universality, because entitlements could not be specified under a system where a pumper's use of water was vulnerable to extraction by a neighbor. Exclusivity did not exist. During periods when pumping was not useful, well owners did not have the option of leasing or selling the water to which they had access. Similarly, transferability was nonexistent. Even if a well owner was paid not to pump water, nothing prevented another landowner from drilling a new well into the aquifer to begin pumping. Thus, a transfer would be rendered meaningless, because the purchaser was not protected from excessive pumping by other users. Finally, there could be no enforceability of a property right for all of the reasons stated above. One pumper could not be prevented from encroaching on another individual's property right.

An owner with a well-defined property right—one that has the four characteristics mentioned above—has a strong incentive to use that resource efficiently, because a decline in the value of that resource represents a financial loss. When well-defined property rights are exchanged, as in a market economy, this exchange facilitates efficiency. Because the seller has the right to prevent the consumer from consuming the product without paying for it, the consumer must pay to receive the product. Given a market price, the consumer will decide how much to purchase by choosing the amount that maximizes individual net benefit.

The state, in its 1996 brief for the Texas Supreme Court during *Barshop v. Medina County Underground Water Conservation District*, recognized the aquifer as a common property resource and called unregulated withdrawals from the aquifer a "tragedy of the commons" in the making, a concept first enunciated by ecologist Garrett Hardin.[66] Common property resources are those not exclusively controlled by a single agent or source. Prior to regulation, landownership was the sole legal requirement for participation in the common property system that characterized the Edwards Aquifer. If access to these resources is not controlled by a single agent or source, the resources will be exploited on a first-come, first-served basis.

Typically, the neoclassical economic approach to solving the problem of overexploitation of common property resources has been to define and enforce property rights through institutional intervention.[67] The government institution protects property rights and manages the resource under goals that promote the public interest. Under a pure rule of capture system for water, property rights, in the economic sense, are an illusion. Existing users are not protected against installation of a well on an adjacent plot of land or withdrawal of water from that well at a rate great enough to lower the water table below the well intakes of surrounding landowners. Indeed, it was this type of unrestricted extraction that ended the rule of capture for oil and gas in Texas, resulting in pooling of underground oil and gas resources. Agricultural interests contended that the regulation of Edwards Aquifer pumping is a taking of private property in *Barshop v. Medina County Underground Water Conservation District*;[68] however, under SB 1477, the regulation and allocation of Edwards water through annual withdrawal permits actually have created quantifiable property rights that were protected under law for the first time.[69]

Policy and Management

The EAA has four primary tasks. The first is to adopt a Critical Period Management Plan (CPMP) for restricting withdrawals during periods when the aquifer level and spring discharge rates are approaching levels adversely affecting endangered species.[70] The second is to issue permits for groundwater pumping based on historical use.[71] The third was to limit total pumping from the aquifer through three staged reductions: to 450,000 ac-ft/yr before December 31, 2007; to 400,000 ac-ft/yr after January 1, 2008; and by December 31, 2012, "the continuous minimum springflows of the Comal Springs and the San Marcos Springs are [to be] maintained to protect endangered and threatened species to the extent required by federal law."[72] The fourth is to manage the aquifer through the development and implementation of a groundwater management plan and the assessment of pumping fees to finance the operation of the EAA.[73] Though the EAA has additional responsibilities, these four were the primary ones assigned by the Texas legislature to resolve the transboundary water disputes associated with the aquifer. The EAA was originally intended to assume these responsibilities on September 1, 1993.[74] However, a series of legal challenges delayed the EAA's operation until a decision by the Texas Supreme Court regarding the constitutionality of its Enabling Act on June 28, 1996.[75]

In addition to the EAA, Senate Bill 1477 created the South Central Texas Water Advisory Committee (SCTWAC) to "advise the EAA Board of Directors on downstream water rights and issues," among other duties.[76] SCTWAC board members are appointed by county judges from counties over the aquifer as well as downstream of it in the watersheds of the Nueces, San Antonio, and Guadalupe Rivers. The EAA board members are elected from districts over the aquifer, and their constituents are primarily aquifer pumpers or those who depend on aquifer pumping for the vast majority of their water.

MANAGING THE EDWARDS AQUIFER

The need for a limit to Edwards Aquifer pumping was recognized by the state of Texas years before the U.S. Congress passed the Endangered Species Act.

Drought Management: Critical Period Management Plan

The Critical Period Management Plan (CPMP) is the set of rules that prescribe how withdrawals from the aquifer will be restricted before spring discharge rates reach critical levels at Comal and San Marcos Springs resulting in take or jeopardy of the listed species and violations of the Endangered Species Act.[77] Historically, drought management plans developed in the San Antonio area have been triggered by an "index well" (J-17) at Fort Sam Houston in Bexar County rather than springflows at Comal and San Marcos. Therefore, the EAA has relied primarily on the levels of three regional groundwater wells (see Figure 5.1) to initiate restrictions on groundwater withdrawals. The levels of the three regional groundwater wells were selected by the EAA to serve as proxies to anticipate when discharge rates for actual springflow approach critical levels.[78]

An examination of the trigger levels used in various drought management plans and the levels for the J-17 index well in San Antonio demonstrates that groundwater withdrawals from the aquifer in many cases would not have been restricted before the onset of take and jeopardy flow levels at Comal Springs. Under the CPMP, the burden for reduced pumping under the critical period restrictions currently is designed to fall disproportionately on municipal water users of the Edwards Aquifer in Bexar, Caldwell, Comal, Hays, and Guadalupe Counties instead of agricultural water users, who are primarily in Medina and Uvalde Counties.[79] The FWS has recommended to the EAA that "trigger levels should be based on springflow rates at Comal (and possibly San Marcos), rather than index well levels."[80] Variations in spring discharges corresponding to the index wells' water levels clearly illustrate the problems with using well levels as proxies for springflow to initiate conservation measures. Even though the overall correlation in some instances is very high between springflow and well levels, such as between the annual means of Comal Springs and J-17, substantial variations exist, which increase as Comal Springs flow declines.[81] It is when the discharge rate is declining that the relationship between the springs and the water level in the index wells becomes critical for triggering conservation measures. Unfortunately, simply raising the index well trigger levels would significantly increase the number of instances when the restrictions would be initiated while spring discharge is safely above the take and jeopardy levels. Although restrictions on aquifer users receive substantial attention during droughts through the CPMP, the overall limit on aquifer pumping is equally critical.

Pumping Limits

The *1961 Texas Water Plan* was the first blueprint for meeting the state's requirements during a repeat of the drought of record. Some major cities that were short of water, such as Dallas and Fort Worth, built reservoirs as the plan recommended. For San Antonio, the plan discouraged overreliance on the Edwards Aquifer and noted that irrigation was depleting groundwater supplies for future municipal use. It claimed that to meet projected 1980 water use, additional regional water supplies would be required, necessitating the construction of new surface-water reservoirs.[82]

The *1968 Texas Water Plan* determined that, based on historical rates of recharge, storage, and hydrologic characteristics, withdrawals should not exceed 400,000 ac-ft annually if water levels in the Edwards Aquifer were to recover following dry periods and a safe yield were to be ensured. The plan also acknowledged that at 400,000 ac-ft, the flow of Comal and San Marcos Springs would be eliminated part of the time. Five years before the ESA was enacted, the Texas Water Development Board (TWDB) stated that the board considered it desirable to maintain some amount of flow from the springs to provide "part of downstream surface water supplies ... as well as enhance the scenic, cultural, and recreational value of the area," among other goals.[83] Thus, justification for guaranteeing minimum flow from the springs initially arose from the generation of economic and aesthetic benefits in New Braunfels, San Marcos, and downstream.

While the *1968 Texas Water Plan* described a potential allocation of the 400,000 ac-ft between pumpers in the three river basins, it also acknowledged that pumping would need to be "reduced somewhat below 400,000 acre-feet annually" to maintain some flow from both Comal and San Marcos Springs.[84] The 400,000-ac-ft figure has survived the test of time. This limit was later revived in a 1992 letter from FWS director Michael Spear to Texas Water Commission (TWC) chairman John Hall, which recommended, "Within 10 years, direct pumpage from the Aquifer shall be reduced by 50,000 acre-feet to 400,000 acre-feet per calendar year."[85] Finally, the 400,000-ac-ft limit was eventually adopted in SB 1477 as the ceiling for annual withdrawals from the Edwards Aquifer after 2007.[86]

The EAA is authorized to achieve the required limits on withdrawals through issued permits or by purchasing and retiring permitted withdrawal rights.[87] Prior to the 2007 amendments to the Edwards Aquifer Authority Act, when the pumping cap was to shrink from 450,000 to 400,000 ac-ft on January 1, 2008, downstream users in the Guadalupe River basin were to contribute 50% of the funds needed to purchase and retire the 50,000 ac-ft.[88] The act also has been interpreted to mean that agricultural irrigators are guaranteed 2 ac-ft of water per acre of irrigated cropland.

After numerous legal challenges, the EAA finished a revised process for issuing permits in 2006. The revised process resulted in 881 regular permits totaling 549,000 ac-ft—some 99,000 ac-ft above the limit specified in SB 1477.[89] The EAA indicated that instead of reducing permitted withdrawals to 450,000 ac-ft before 2008 and 400,000 ac-ft after 2008, the authority would seek to raise authorized pumping limits to 549,000 ac-ft annually through an amendment to SB 1477.[90]

This possibility prompted a request by State Representative Patrick Rose in 2006 that the TWDB prepare an analysis of various scenarios being proposed by the EAA to limit pumping during a repeat of the drought of record.[91] The analysis was completed in 2007 and determined that even if pumping levels were limited to 340,000 ac-ft/yr during a repeat of the drought of record with current drought management restrictions fully enforced, Comal Springs would cease to flow for 25 to 30 months.[92] During the actual drought of record, Comal Springs ceased to flow for less than 5 months.[93] Cessation of flow from Comal Springs for any period of time would jeopardize the existence of endangered species at the springs, as well as the hydrology, ecology, and economy of most of the Guadalupe River basin and San Antonio Bay.

Major Attempts to Develop Alternative Water Supplies for San Antonio

All attempts to develop a major source of water to supplement or offset San Antonio's use of the Edwards Aquifer have failed thus far. In 1991 and again in 1994, in special referendums, the citizens of San Antonio voted not to complete the nearby Applewhite Reservoir, under construction on the Medina River southwest of the city.[94] This project was one in a series of supplemental water supplies turned down by San Antonio, including the City Council's rejection of the purchase of 30,000 ac-ft of water from nearby Canyon Reservoir on the Guadalupe River in 1976. In the 1950s and 1960s, the Guadalupe-Blanco River Authority (GBRA) and San Antonio had fought over control of the Canyon Reservoir project, with the GBRA the winner in the Texas Supreme Court. Prior to the 1950s, the city refused participation in the U.S. Army Corps of Engineers Goliad Reservoir project on the San Antonio River. Still earlier, before World War II, the San Antonio city fathers declined an offer to buy up the water rights in the San Antonio River watershed from Medina Lake.[95] In 1997, the legislature hoped that the Senate Bill 1 water planning process might finally produce a major alternative water supply to San Antonio's nearly sole reliance on the aquifer.

In 2001, the GBRA, San Antonio Water System (SAWS), and San Antonio River Authority (SARA) signed an agreement to bring a minimum of 70,000 ac-ft/yr of surface water to San Antonio in what was known as the Lower Guadalupe Water Supply Project (LGWSP). This surface water, along with lesser amounts of groundwater from the Gulf Coast Aquifer, was scheduled to reach

San Antonio beginning in 2010, thereby relieving some of the demand on the Edwards Aquifer and providing the city with a supplemental supply of surface water, as well as some protection for springflow from Comal and San Marcos Springs, instream flows in the Guadalupe River, and bay and estuary inflows to San Antonio Bay.[96] However, on August 16, 2005, the SAWS Board of Directors officially withdrew from the LGWSP.[97] A similar joint project by the San Antonio Water System and Lower Colorado River Authority, the SAWS-LCRA Project, was canceled in 2009, leaving San Antonio without any significant alternative supply of water in the foreseeable future.[98]

LEGISLATIVE CHANGES IN DIRECTION

The most significant Edwards Aquifer legislation since 1993 was passed during the 80th Session of the Texas legislature on May 28, 2007, and signed into law by Governor Rick Perry on June 16. Senate Bill 3, Article 12, makes two major changes to the management of the Edwards Aquifer. First, the legislation lifts the Edwards Aquifer pumping cap to 572,000 ac-ft/yr from the 400,000 ac-ft/yr limit that was required by January 1, 2008.[99] Second, the EAA is required, with the assistance of Texas A&M University (TAMU), to develop a Recovery Implementation Program (RIP) through a facilitated, consensus-based process that involves input from the FWS, other appropriate federal agencies, and all interested stakeholders. This article also placed in statute pumping restrictions that the EAA must enforce, although these restrictions would allow Comal Springs to go dry for up to three years during a repeat of the drought of record.

The Edwards Aquifer Recovery Implementation Program

In mid-2007, EAA and TAMU began to fulfill the second requirement by developing the Edwards Aquifer Recovery Implementation Program (EARIP), a voluntary, multi-stakeholder initiative that seeks to balance water use and development with the recovery of federally listed species. Because of the diversity of issues and level of conflict often associated with water issues, RIPs use a long-term, interdisciplinary approach that incorporates policy formation, scientific research, habitat restoration, education, and other activities as defined by the participants. The implementation time frame for existing programs ranges from 15 to 50 years and may be extended if necessary. Formation of a RIP requires that the stakeholders participating in the program develop a comprehensive document that outlines the program goals, activities, timelines, measurements of success, and roles of the participants.[100] Development of the program document—in the case of the EARIP, a Habitat Conservation Plan (HCP)—can take years to decades. Once the program document is finalized, stakeholders interested in participating in program implementation sign a cooperative agreement to implement the activities outlined in the program document. The EARIP is unique for several reasons, which include a number of firsts for RIPs under the Endangered Species Act:

- No federal nexus is associated with the proposed action, and thus no Section 7 consultation with the FWS (federal agencies must review their actions and determine whether these actions may affect federally listed and proposed species or proposed or designated critical habitat) is associated with this RIP.
- The EARIP is fully based on the development of an HCP.
- Federal water management is not part of the proposed action.
- The focus is management of an aquifer.
- The program was mandated by state legislation.
- The EARIP was developed in a state consisting predominantly of private lands.
- A coastal/estuary ecosystem is included in the project area.[101]

The initial 21-member Steering Committee consisted of representatives of state and regional surface-water and groundwater agencies and interests, as well as municipalities, industries, agriculture, and the public. The committee voted to add more members, expanding to 26 in 2008.

As called for in the Edwards Aquifer Authority Enabling Act, an Edwards Aquifer Area Expert Science Subcommittee was appointed, with the following responsibilities:

- evaluate designating a separate San Marcos Pool, which would relieve aquifer pumpers in other counties from pumping restrictions, particularly Bexar County, where San Antonio is located;
- evaluate the necessity to maintain minimum springflows;
- evaluate whether adjustments to flow triggers for San Marcos Springs should be made;
- analyze species requirements for springflow and aquifer levels as a function of recharge and withdrawal levels;
- develop recommendations for withdrawal reductions for critical period management in a collaborative process designed to achieve consensus; and
- submit the recommendations to the Steering Committee and all other stakeholders involved in the EARIP.[102]

The act also calls for the EAA, Texas Commission on Environmental Quality, Texas Parks and Wildlife Department, Texas Water Development Board, and U.S. Fish and Wildlife Service to approve and execute the RIP document (the HCP) by September 30, 2012.[103] The EAA Board is given the final say over whether to implement the recommendations coming out of the RIP process.[104] Finally, the EAA will provide a written report to the governor, lieutenant governor, and Speaker of the Texas House of Representatives describing the actions taken in response to each recommendation of the HCP and explaining why any were not implemented.[105] The agreement takes effect on December 31, 2012.[106]

The EARIP has been making progress. Stakeholders have met monthly since mid-2007, and all procedural tasks were completed in less than six months, including hiring Robert Gulley as program manager. Thirty-seven stakeholders have signed a memorandum of agreement to work cooperatively to find solutions to balance the water demands of the regional population and endangered species.[107] The process is now focusing on the difficult substantive questions that have vexed the region for decades. At a July 17, 2008, hearing of a Texas Senate Natural Resources Subcommittee, Gulley testified that the EARIP has made a tremendous amount of progress and thus far has worked collaboratively and cooperatively.[108]

On November 13, 2008, the Edwards Aquifer Area Expert Science Subcommittee completed its deliberations on the first three areas of responsibility listed above. Regarding the first charge, to evaluate whether designation of a separate San Marcos Pool was feasible, the subcommittee concluded that it did not believe that sufficient data existed to support designation or define the boundaries of such a pool. On the second charge, the subcommittee found that minimum springflows were necessary "for the survival and recovery of each species listed under the Endangered Species Act." Finally, on the third charge, the subcommittee concluded that adjustments should not be made to flow triggers for San Marcos Springs.[109]

On July 10, 2009, the EARIP Steering Committee decided to focus on a habitat conservation plan as its comprehensive document. Prior to 1982, nonfederal parties faced penalties under the Endangered Species Act when their otherwise legal activities resulted in the take of a species. In that year, Congress amended the ESA to allow the taking of federally listed species when the take is the inadvertent result of a legal activity by obtaining an incidental take permit (ITP) under Section 10(a) of the act.[110] With regard to the Edwards Aquifer, an ITP would allow withdrawals that may cause the take of listed species at Comal or San Marcos Springs to continue until the jeopardy spring discharge levels are reached.[111] To secure a permit, protective measures are devised to prevent springflows from declining below specific levels, in turn preventing jeopardy for the species of concern. Development of a Habitat Conservation Plan (HCP) is required for an ITP.[112] A regional Edwards Aquifer HCP will be used to secure a multiyear permit authorizing incidental takes by those entities and individuals who sign the application.

What Lies Ahead for the EARIP

Despite the progress made thus far, the EARIP will face many challenges before it is done. Determining the geographic scope of the EARIP will be a key issue. Will the process be confined to the species found in and near the Edwards Aquifer springs, or will it include areas downstream of the springs where other endangered species can be found, such as San Antonio Bay? The requirement to maintain continuous minimum natural springflows throughout a repeat of the drought of record will be another key issue for the EARIP. Near-zero flows for any extended period of time could, and likely would, lead to the extinction of one or more listed species, and of these, only the fountain darter and Texas wild-rice will reliably reproduce in captivity.[113] Thus, far, two very different views on the central issue of the necessity of maintaining continuous minimum springflows have been expressed by EARIP participants, demonstrating the challenges that await the stakeholders:

> San Antonio Water System: "It is our position that continuous minimum springflows are not essential to the survival of the species, as the species have survived periods of extreme drought, and current aquifer management strategies have ensured more than adequate springflow for the species based on what have been recognized as extremely conservative 'take' and 'jeopardy' numbers for the Edwards Aquifer species. A science-based review of the data on springflow will likely result in a determination of no minimum springflow requirements for the species."[114]

> Texas Parks and Wildlife Department: "TPWD staff believes it is imperative to maintain at least minimum springflows during periods of drought, and preferably a flow regime that mimics natural hydrologic conditions at all times, in order to maintain the federally listed species in their native habitat. It should go without saying that aquatic organisms need water."[115]

Through SB 3, the Texas legislature has created a model for implementing federal RIPs that other states likely will replicate to address intractable ESA disputes by constraining the federal process in terms of timing, issues to be addressed, and governance.

CONCLUSIONS

The Edwards Aquifer is a common-pool resource that, after 10 years of regulation by the EAA, is still undergoing a difficult transition to a regulated resource at a time when the aquifer is unable to satisfy all desires for domestic, municipal, industrial, commercial, agricultural, recreational, and environmental water. The continuing conflict over the Edwards Aquifer began in the 1950s during the Texas drought of record, years before the Endangered Species Act became law in 1973 (for a timeline of events, see the appendix at the end of this chapter). The struggle pits urban versus rural culture and economics; agricultural interests in Bexar, Medina, and Uvalde Counties versus municipal, recreational, and industrial interests in San Antonio; and all of these versus various interests in the spring communities and downstream in the Guadalupe River basin. The inexpensive nature of the Edwards groundwater compared to other major water supply alternatives continues to be the primary driver of San Antonio water policy, and with the EAA's elected board representing San Antonio and aquifer water right holders and pumpers, it is the authority's driving policy as well.

If Comal and San Marcos Springs are to continue flowing on a permanent basis, measures are required that include the following: conservation of Edwards Aquifer water; adoption of a regional drought management plan that will preserve springflow in a repeat of the drought of record;[116] continued development of an efficient market for trading Edwards Aquifer water rights; development of significant amounts of additional surface and non-Edwards groundwater supplies; and development of a regional habitat conservation plan to obtain an ESA Section 10(a) incidental take permit. The use of Edwards Aquifer groundwater has become much more efficient with regulation. However, the development of any significant alternative water supplies has yet to materialize more than 50 years after Comal Springs ceased to flow in 1956, despite multiple state water plans and a federal court judgment directing otherwise. The region probably will continue the move to regional water management, as its major aquifers and rivers are closely interlinked. However, the political hurdles that the region faces to achieve this goal are likely to be surmounted only in the midst of crisis resulting from prolonged drought. San Antonio cannot reduce its reliance on the Edwards Aquifer through conservation alone if the economy and population of the city are to continue to grow. It will also require new supplies, most of which will require cooperation from other regional interests that control access to the available sources of water.

No one knows when a repeat of the drought of record will begin. Droughts have driven the development of Texas water management policies, programs, and law. Accordingly, major water legislation and litigation have followed droughts. The continuing conflict over the Edwards Aquifer has been the catalyst for major changes in Texas water policy over the past half century. Left unaddressed, rapid population growth in the Edwards region without significant sources of additional water may well result in a future crisis at Comal or San Marcos Springs, or both, even if water use continues to become more efficient. The remedy is apt to lead once again to key changes in Texas water policy that affect surface-water and groundwater management throughout the state.

APPENDIX: CHRONOLOGY OF MAJOR EVENTS

Prior to 1884	Comal and San Marcos Springs are among the most prolific U.S. springs and have continuous discharge at all times, even during major droughts. Unique spring-dependent species evolve in these ecosystems.
1884	The first irrigation well is completed in Bexar County.
1900	Aquifer withdrawals reach approximately 30,000 ac-ft/yr.
1904	Rule of capture is adopted as the state's groundwater law by the Texas Supreme Court in *Houston & Ry. v. East.*
1947–1957	The drought of record. Comal Springs ceases to flow for 144 days in 1956. San Marcos Springs drops to 46 cfs. Some report that bad-quality water has migrated into the aquifer's freshwater zone. Beverly Lodges index well (later replaced by J-17) drops to 612 feet mean sea level.
1949	The state authorizes voluntary creation of underground water conservation districts.
1955	The Texas Supreme Court rules San Antonio has water supply problem, requiring new water sources, in *Board of Water Engineers v. San Antonio.* The legislature attempts to form the Edwards Underground Water District (EUWD).
1959	The legislature creates the EUWD to protect and preserve the aquifer.

1961	The Board of Water Engineers publishes the first Texas Water Plan, which recommends a surface-water supply for San Antonio.
1967	The U.S. Fish and Wildlife Service (FWS) lists the Texas blind salamander as endangered.
1968	The Texas Water Development Board updates the Texas Water Plan, which recommends a 400,000 ac-ft/yr Edwards Aquifer withdrawal limit.
1970	The Texas Water Quality Board orders aquifer water quality protection. The fountain darter is listed as endangered.
1972–1984	The EUWD builds four small recharge dams over the Edwards Aquifer.
1973	The modern Endangered Species Act (ESA) becomes law.
1978	Texas wild-rice is listed as endangered and the San Marcos salamander is listed as threatened. The ESA is amended to require the preparation of recovery plans.
1980	The San Marcos gambusia is listed as endangered. Critical habitat is designated at San Marcos Springs.
1980–1990	Aquifer withdrawals increase significantly after the drought of record, reaching 542,500 ac-ft/yr in 1989.
1985	The San Marcos Recovery Plan is adopted by FWS.
January 1989	Uvalde and Medina Counties secede from the EUWD over a disagreement about limiting withdrawals.
June 15, 1989	The Guadalupe-Blanco River Authority (GBRA) issues a notice of intent to sue for ESA violations. It files suit in Hays County District Court to have the aquifer declared an underground river owned by state. (The Hays County case is still pending as of 2010.)
April 12, 1990	The Sierra Club issues a notice of intent to sue for violation of the ESA.
1991	Living Waters Artesian Springs catfish farm opens southwest of San Antonio, using up to 40 million gallons/day from the aquifer.
May 16, 1991	The Sierra Club files a lawsuit in U.S. District Court, *Sierra Club v. Lujan* (later *Babbitt*). The GBRA, San Antonio, and others intervene. The suit alleges that Interior and FWS failed to protect aquifer-dependent endangered species, violating the ESA. The plaintiffs ask the court to order FWS to determine the minimum spring discharge required at Comal and San Marcos Springs to avoid species takes and jeopardy.
April 15, 1992	The Texas Water Commission (TWC) declares the aquifer an underground stream and therefore state water. It adopts emergency rules and initiates rulemaking proceedings. Pumpers would be allocated 450,000 ac-ft/yr.
August 1992	A Travis County Court voids TWC's declaration that aquifer is an underground river. The aquifer rules are voided.

November 16–19, 1992	Trial in *Sierra Club v. Babbitt* before Judge Bunton in Midland.
February 1, 1993	Judge Bunton finds in favor of Sierra Club and others that "firm yield" of aquifer is about 200,000 ac-ft/yr, far below the 542,500 ac-ft/yr withdrawn in 1989. He rules that if withdrawals continue unabated, spring discharge will diminish and listed species will be taken, violating ESA. Texas Natural Resource Conservation Commission (TNRCC) is directed to devise plan to limit withdrawals and preserve springs during repeat of drought of record. Legislature is given until May 31 to enact TNRCC plan or judge will allow plaintiffs to seek federal judicial control of aquifer. FWS is ordered to determine species take and jeopardy spring discharge levels.
April 15, 1993	Pursuant to Judge Bunton's order, the FWS determines that takes begin when Comal Springs discharge declines to 200 cfs and jeopardy occurs below 150 cfs. For San Marcos Springs, jeopardy begins when discharge declines to 100 cfs.
May 30, 1993	The 73rd Legislature creates the Edwards Aquifer Authority (EAA) to regulate groundwater use. The EUWD is abolished.
September 1, 1993	Senate Bill (SB) 1477 is to take effect but is delayed while the U.S. Department of Justice (DOJ) decides whether abolition of the EUWD's elected board and substitution of an appointed board violates the Voting Rights Act.
November 19, 1993	DOJ rules that SB 1477 does not meet the requirements of the Voting Rights Act because it would abolish an elected board (EUWD), replacing it with an appointed one (EAA).
October 27, 1995	Medina County judge Pennington rules that SB 1477 is unconstitutional in *Barshop v. MCUWCD*.
June 28, 1996	A unanimous Texas Supreme Court reverses *Barshop v. MCUWCD*, finding SB 1477 constitutional.
August 23, 1996	*Sierra Club v. San Antonio*, special master, Todd Votteler, completes emergency drought plan for Judge Bunton. Judge Bunton declares a water emergency.
September 11, 1996	Judge Bunton's August 23, 1996, order is stayed by the U.S. 5th Circuit Court of Appeals.
March 13, 2007	The EAA Board raises withdrawal cap to 549,000 ac-ft/yr through 2008.
May 28, 2007	Legislature approves Senate Bill 3 and House Bill 3 in final hours of session. Identical language allows 572,000 ac-ft/yr of withdrawals. The Edwards Aquifer Recovery Implementation Program is mandated.
July 17, 2007	FWS designates critical habitat at Comal Springs for three invertebrates: the Comal Springs riffle beetle, Comal Springs dryopid beetle, and Peck's cave amphipod.

2007–present   A lawsuit is filed by the Center for Biological Diversity over former Interior Department official Julie MacDonald's interference in the designation of critical habitat at Comal Springs for the three invertebrates.

NOTES

1. Jerry Needham, Aquifer/Drought Rules: Law Sets New Limits on Water, *San Antonio Express-News*, May 27, 2008, 1B. Other sources indicate that the Edwards Aquifer provides 93% of San Antonio's water supply. Janelle Okorie, Testimony before the Edwards Aquifer Authority Legislative Oversight Committee, Austin, TX, October 5, 2006.

2. Texas Const. art. I, § 17.

3. U.S. Const. amend. V.

4. Endangered Species Act of 1973 (ESA), 16 U.S.C. § 1544 (1994).

5. Edwards Aquifer Authority (EAA), History, http://www.edwardsaquifer.org/pages/history.htm (accessed March 18, 2007; site now discontinued).

6. John J. Vandertulip, ed., *A Plan for Meeting the 1980 Water Requirements of Texas* (Austin, TX: Texas Board of Water Engineers, 1961).

7. Laura A. Wimberley, Reluctant Conservationists, Water Scarcity and Regional Interdependence: Central Texans and the "Great Drought" (unpublished manuscript presented at the Southwest Social Science Association Annual Meeting, New Orleans, LA, March 28, 1997; on file with author).

8. FR 58344 12/16/75; and 42 U.S.C. §§ 300f–300j (1994).

9. Steven Johnson and Geary Schindel, *Evaluation of the Option to Designate a Separate San Marcos Pool for Critical Period Management*, Report 08-01 (San Antonio, TX: EAA, February 2008); Larry Land and Paula Jo Lemonds, Technical Memo: Hydrologic Connection of the Edwards Aquifer between San Marcos Springs and Barton Springs (Austin, TX: HDR Engineering, October 17, 2009); Larry Land, Evaluation of Hydrologic Connection between San Marcos Springs and Barton Springs through the Edwards Aquifer (Austin, TX: HDR Engineering, April 2010).

10. U.S. Geological Survey, Memorandum, 1, 1997 (on file with author); Barton Springs/Edwards Aquifer Conservation District website, Aquifer Science–Geology, http://www.bseacd.org/geology.html (accessed June 22, 2008; site now discontinued).

11. EAA, *The Edwards Aquifer: A Texas Treasure* (San Antonio, TX: EAA, n.d.).

12. Todd H. Votteler, The Little Fish That Roared: The Endangered Species Act, State Groundwater Law, and Private Rights Collide over the Texas Edwards Aquifer. *Environmental Law* 29 (1998): 845, 847–48, available at http://www.edwardsaquifer.net/votteler.html (accessed June 16, 2010).

13. Gunnar Brune, *The Springs of Texas*, vol. 1 (College Station, TX: Texas A&M University Press, 1981).

14. U.S. Geological Survey (USGS), *Recharge to and Discharge from the Edwards Aquifer in the San Antonio Area, Texas* (San Antonio, TX: USGS, 1999), 3.

15. Mike Thuss (chief executive officer of the San Antonio Water System), interview by Todd Votteler, August 25, 1999.

16. Texas Water Development Board (TWDB), *Water for Texas 2007*, vol. II, Document GP-8-1 (Austin, TX: TWDB, January), 128.

17. San Antonio Water System, Water Statistics: Year Ending December 31, 1992 (San Antonio, TX, 1993), B-42; Jaie Avila, Investigation: Where Did All That Water Go? WOAI

News 4, July 29, 2008. According to Avila, in 2007, the Bexar Metropolitan Water District lost 22% of its water in transmission.

18. Karen Guz, Conservation Success and Challenges (presented at From Policy to Reality: Advanced Urban Water Conservation in Texas conference, Austin, TX, April 18, 2008), 1.

19. In addition to Guadalupe River discharge, total bay and estuary inflows consist of San Antonio River discharge and inflows from ungauged coastal basins. See also D.J. Ockerman and R.N. Slattery, *Streamflow Conditions in the Guadalupe River Basin, South-Central Texas, Water Years 1987–2006: An Assessment of Streamflow Gains and Losses and Relative Contribution of Major Springs to Streamflow*, U.S. Geological Survey Scientific Investigations Report 5165 (Austin, TX: USGS, 2008).

20. USGS, *Recharge to and Discharge from the Edwards Aquifer* (1999).

21. USGS, Ground-Water Storage in the Edwards Aquifer, San Antonio Area, Texas, available at http://tx.usgs.gov/reports/dist/dist-1996-01/dist-1996-01.pdf (accessed June 16, 2010).

22. W.B. Klemt, T.R. Knowles, G.R. Elder, and T.W. Sieh, *Ground-Water Resources and Model Applications for the Edwards (Balcones Fault Zone) Aquifer in the San Antonio Region* (Austin, TX: Department of Water Resources, 1979) http://www.twdb.state.tx.us/publications/reports/GroundWaterReports/GWReports/R239/R239.pdf (accessed June 16, 2010).

23. B.D. Jones, National Water Summary, 1988–89: Hydrologic Events and Floods and Droughts, USGS Water-Supply Paper 2375, 1991.

24. Texas Administrative Code, § 357.2(2).

25. Ibid., §§ 357.7(a)(3), 357.5(e)(2).

26. Water Demand/Drought Management Technical Advisory Committee of the Consensus State Water Plan, Potential Impacts of Drought in Texas (paper presented at Planning for the Next Drought: A National Drought Mitigation Center Workshop, 1998).

27. TWDB, *Water for Texas: A Consensus-Based Update to the State Water Plan*, Technical app. GP-6-2, 2-36 (1997). Other sources use 1950 to 1957.

28. Todd H. Votteler, Water from a Stone: The Limits of the Sustainable Development of the Texas Edwards Aquifer (PhD diss., Southwest Texas State University, 2000), 209; Glenn Longley, The Relationship between Long Term Climate Change and Edwards Aquifer Levels, with an Emphasis on Droughts and Spring Flows (paper presented at the 24th Water for Texas Conference, Austin, TX, January 1995).

29. Larry Land and Tricia Sebes, Technical Memo: Natural Springflow from the Edwards Aquifer—San Antonio Segment (Austin, TX: HDR Engineering, December 2008), 3.

30. Votteler, Water from a Stone, 73; Tables 5.1 and 5.2; D.S. Brown and J.T. Patton, *Recharge to and Discharge from the Edwards Aquifer in the San Antonio Area, Texas, 1995* (San Antonio, TX: USGS, 1996); and USGS, *Recharge to and Discharge from the Edwards Aquifer in the San Antonio Area, Texas, 1934–2007* (San Antonio, TX: USGS, 2008). USGS average annual recharge from 1934 to 1995 was 674,200 ac-ft/yr; however, recent high-recharge years have skewed average annual recharge to 732,400 ac-ft/yr during the period 1934 to 2007.

31. Jones, National Water Summary, 1988–89, 518; TWDB, *Water for Texas Update*, 2-36; David Stahle and Malcolm Cleaveland, Texas Drought History Reconstructed and Analyzed from 1698 to 1980, *Journal of Climate* 1 (1988): 72; Malcolm K. Cleaveland, *Extended Chronology of Drought in the San Antonio Area* (Fayetteville, AR: Tree-Ring Laboratory, University of Arkansas, 2006), 15–16.

32. Votteler, Water from a Stone.

33. H.A. Loaiciga, D.R. Maidment, and J.B. Valdes, Climate-Change Impacts in a Regional Karst Aquifer, Texas USA, *Journal of Hydrology* 227 (2000): 173–74, 187.

34. C. Chen, D. Gillig, and B. McCarl, Effects of Climatic Change on a Water Dependent Regional Economy: A Study of the Texas Edwards Aquifer, *Climate Change* 49 (2001): 397–409.

35. R. Seager, M.F. Ting, I.M. Held, Y. Kushnir, J. Lu, G. Vecchi, H.-P. Huang, N. Harnik, A. Leetmaa, N.-C. Lau, C. Li, J. Velez, and N. Naik, Model Projections of an Imminent Transition to a More Arid Climate in Southwestern North America, 2007, 10/1126/*Science*.1139601, http://www.ldeo.columbia.edu/cicar/documents/Sc_Express_Model_Predictionsv2.pdf (accessed June 16, 2010).

36. Glenn Longley, The Edwards Aquifer: Earth's Most Diverse Groundwater Ecosystem? *International Journal of Speleology* 11 (1981): 123, 127. See also Rick Illgner, The Edwards Aquifer: Political Prisoner (paper presented at the 89th Annual Meeting of the Association of American Geographers, Atlanta, GA, April 1993, on file with author).

37. Glenn Longley and Henry Karnei, Jr, *Status of Satan Eurystomus Hubbs and Bailey, the Widemouth Blindcat*, vol. 6 (San Marcos, TX: Southwest Texas State University, 1978).

38. San Marcos/Comal Recovery Team, U.S. Fish and Wildlife Service (FWS), *San Marcos and Comal Springs and Associated Aquatic Ecosystems (Revised) Recovery Plan* (Austin, TX: FWS, 1996), 121.

39. Mara Alexander and Catherine Phillips, *Endangered and Threatened Species of the Edwards Aquifer* (paper presented at the Edwards Aquifer Recovery Implementation Program, San Antonio, TX, April 5, 2007), 21.

40. Endangered Species Act, 16 U.S.C. § 1532 (19) (2001).

41. *Sierra Club v. Lujan*, No. MO-91-CA-069, 1993 WL 151353, 6 (W.D. Tex. Feb. 1, 1993).

42. *Notice of Filing of Springflow Determinations Regarding "Take" of Endangered and Threatened Species, submitted by Charles R. Shockley on behalf of U.S. Department of Justice, Environmental and Natural Resources Division following Sierra Club v. Lujan*, No. MO-91-CA-069, 1993 WL 151353 (W.D. Tex. Feb. 1, 1993) (on file with author); *Notice of Filing of Springflow Determinations Regarding Survival and Recovery and Critical Habitat of Endangered and Threatened Species, submitted by Charles R. Shockley on behalf of U.S. Department of Justice, Environmental and Natural Resources Division following Sierra Club v. Lujan*, No. MO-91-CA-069, 1993 WL 151353 (W.D. Tex. Jun. 15, 1993) (on file with author).

43. San Marcos/Comal Recovery Team, *Revised Recovery Plan; Sierra Club v. Lujan*.

44. Ronald A. Kaiser, *Handbook of Texas Water Law: Problems and Needs*, Texas Water Resources Institute, No. 32 (College Station, TX: Texas A&M University, 1987).

45. See *Houston & T.C. Ry. Co. v. East*, 81 S.W. 279 (Tex. 1904).

46. Kaiser, *Handbook of Texas Water Law.*

47. House Research Organization, Texas House of Representatives, *Regulating the Edwards Aquifer: A Status Report*, 73-8 (1994), 19.

48. Rick Illgner, The Edwards Aquifer: Political Prisoner (paper presented at the 89th Annual Meeting of the Association of American Geographers, Atlanta, GA, April 1993, on file with author), 4.3.

49. Kaiser, *Handbook of Texas Water Law.*

50. Edwards Aquifer Authority Enabling Act, ch. 626, 1993 Tex. Gen. Laws 2355.

51. For more details on *Sierra Club v. Babbitt*, as well as other Edwards Aquifer cases, see Votteler, Little Fish That Roared.

52. *Sierra Club v. Lujan.*

53. See generally ibid.

54. Lucius D. Bunton III, *A Bit of Bunton: Memoirs by Lucius D. Bunton III* (Odessa, TX: Lucius D. Bunton, 1999), 310–11.

55. *Sierra Club v. Lujan.*

56. Stuart Henry (attorney for the Sierra Club), interview by Todd Votteler, August 25, 1999.

57. *Sierra Club v. Lujan*, at 29; *Finding 196, Amended Findings of Fact and Conclusions of Law, Sierra Club v. Lujan*, May 26, 1993.

58. See EAA Enabling Act, ch. 626, 1993 Tex. Gen. Laws 2355.

59. House Research Org., *Regulating the Edwards Aquifer*. Attempts to create the EUWD had failed in the legislature during the 1955 and 1957 sessions.

60. EAA Enabling Act, ch. 626, § 1.14.

61. Ibid., §§ 1.14(b)–(c).

62. Ibid., § 1.14(h).

63. Ibid., §§ 1.14(a)–(b).

64. Bob Stallman, Endangered Species Task Force (statement at House Resources Field Hearing in Boerne, Texas, March 20, 1995).

65. Tom Teitenberg, *Environmental and Natural Resources Economics*, 4th ed. (New York: Addison Wesley, 1996).

66. *Brief for Appellant at 8–9, Barshop v. Medina County Underground Water Conservation Dist.*, 925 S.W.2d 618 (Tex. 1996) (No. 95-0881), quoting Garrett Hardin, The Tragedy of the Commons, *Science* 162 (1968): 1243–44.

67. G.A. Tobin et al., Water Resources, in *Geography in America*, ed. Gary L. Gaile and Cort J. Wilmot, 127 (New York: Oxford University Press, 1989).

68. See 925 S.W.2d 618, 618 (Tex. 1996).

69. EAA Enabling Act, ch. 626, 1993 Tex. Gen. Laws 2355 § 1.14.

70. Ibid., § 1.26(a).

71. Ibid., § 1.16.

72. Ibid., § 1.14(b),(c),(h).

73. Ibid., §§ 1.25–1.29.

74. Ibid., § 4.02.

75. See *Barshop v. Medina County Underground Water Conservation Dist.*, 925 S.W.2d 618 (Tex. 1996).

76. Naismith Engineering, South Central Texas Water Advisory Commission: Report of the Effectiveness of the Edwards Aquifer Authority (Corpus Christi, TX: Naismith Engineering, 2000), 1.

77. EAA Enabling Act, ch. 626, 1993 Tex. Gen. Laws 2355, § 1.26.

78. See Votteler, Water from a Stone.

79. Ibid.

80. David Frederick, supervisor, FWS, letter to Greg Ellis, general manager of the EAA, 1998 (on file with author).

81. Votteler, Water from a Stone.

82. Vandertulip, *Plan for Meeting 1980 Water Requirements*.

83. TWDB, *Texas Water Plan 1968*, I-15 (Austin, TX: State of Texas, 1968).

84. Ibid.

85. Michael Spear, regional director, FWS, letter to John Hall, chairman, Texas Water Commission, 1992.

86. Act of May 30, 1993, 73rd Leg., R.S., ch. 626, 1993 Tex. Gen. Laws 2355, § 1.14(c).

87. EAA Enabling Act, ch. 626, 1993 Tex. Gen. Laws 2355 §§ 1.16, 1.22a.

88. Ibid., § 1.29(d). Domestic and livestock pumping was excluded from the 450,000 and 400,000 ac-ft caps.

89. South Central Texas Water Advisory Committee (SCTWAC), *Fifth Biennial Report on the Effectiveness of the Edwards Aquifer Authority, Executive Summary* (Austin, TX: SCTWAC, 2006), sec. 2, 6.

90. EAA, *2007 Legislative Agenda* (San Antonio, TX: September 12, 2006).

91. Patrick Rose, state representative, to Kevin Ward, executive administrator of the TWDB, and Glenn Shankle, executive director of the TCEQ, November 22, 2006.

92. Shirley Wade, Robert Mace, and Andrew Donnelly, GAM Run 06-33a (Austin, TX: Texas Water Development Board, February 12, 2007).

93. Votteler, Water from a Stone, 1.

94. Votteler, Little Fish That Roared.

95. Todd H. Votteler, Raiders of the Lost Aquifer? Or, the Beginning of the End to Fifty Years of Conflict over the Texas Edwards Aquifer, *Tulane Environmental Law Journal* 15 (2002): 299, available at http://www.edwardsaquifer.net/votteler.html (accessed June 16, 2010).

96. GBRA Moves to Secure Water Supply to Region, *Guadalupe-Blanco River Authority News*, November 28, 2001, 2; Water Supply and Delivery Agreement among Guadalupe-Blanco River Authority, San Antonio Water System and San Antonio River Authority, § 5.2.3, May 10, 2001 (on file with author).

97. Jerry Needham, SAWS Won't Dip Deeper into the Aquifer: 50-Year Plan Also Pulls Plug on Pair of Pricey Pipeline Projects, *San Antonio Express-News*, August 17, 2009, 1A.

98. Asher Price, LCRA Backs Off San Antonio Water-Sharing Deal, *Austin American-Statesman*, April 10, 2009, 1A.

99. In 2008, Robert Puente, coauthor of House Bill 3 and sponsor of Senate Bill 3, retired from the Texas House of Representatives, where he was chairman of the Natural Resources Committee. After returning to his district in San Antonio, he was named the chief executive officer of the San Antonio Water System. San Antonio Water System, Robert R. Puente Named New Leader at SAWS, press release, November 24, 2008. In 2008, EAA chairman Doug Miller, from New Braunfels, was elected representative of District 73 of the Texas House. Chris Cobb, Miller Elected to State House, *Herald-Zeitung*, November 5, 2008, 1A.

100. Joy Nicholopoulos, *Balancing Waters Needs in the Edwards Aquifer: The Collaborative Approach of Recovery Implementation Programs* (paper presented at the Introductory Meeting of the Proposed Edwards Aquifer Recovery Implementation Program, Aquarena Center, San Marcos, TX, February 16, 2007); and Texas A&M Institute of Renewable Natural Resources, Edwards Aquifer Recovery Implementation Program, Overview, http://irnr.tamu.edu/earip/index.cfm (accessed October 31, 2007; site now discontinued).

101. Anna M. Munoz, e-mail message to author, March 25, 2008.

102. EAA Enabling Act, ch. 626, 1993 Tex. Gen. as amended in 2007, § 1.26A(j, k).

103. Ibid., § 1.26A(d)(3).

104. Todd H. Votteler, Legislature Passes and Governor Signs Major Water Legislation, *GBRA River Run* (Summer 2007): 7.

105. EAA Enabling Act, § 1.26(a).

106. Ibid., § 1.26A(d)(3).

107. Edwards Aquifer Recovery Implementation Program, Edwards Aquifer RIP Testifies of Its Success to Senate Legislative Committee, news release, July 18, 2008.

108. Written testimony of Robert Gulley, project manager for the Edwards Aquifer Recovery Implementation Program, before the Texas Senate Natural Resources Subcommittee on Regional Water Quality and EAA RIP, July 17, 2008.

109. Edwards Aquifer Area Expert Science Subcommittee, *Evaluation of Designating a San Marcos Pool, Maintaining Minimum Spring Flows at Comal and San Marcos Springs, and Adjusting the Critical Period Management Triggers for San Marcos Springs*, Report to the Steering Committee for the Edwards Aquifer Recovery Implementation Program, November 13, 2008, v–vii, http://earip.tamu.edu/Science/Documents.aspx (accessed June 16, 2010).

110. Endangered Species Act of 1973, 16 U.S.C. § 1539(a)(1)(B) (1997).

111. Joe G. Moore, Jr, and Todd H. Votteler, Draft Habitat Conservation Plan for the Edwards Aquifer (Balcones Fault Zone—San Antonio Region), prepared for the Honorable Lucius D. Burton III, June 23, 1995 (on file with author).

112. U.S.C. § 1539(a)(2)(A).

113. Thomas Brandt, Refugia (paper presented at the Edwards Aquifer Recovery Implementation Program Science Subcommittee, Texas River Systems Institute, May 19, 2008).

114. Calvin Finch, Edwards Aquifer Recovery Implementation Program, Senate Bill 3, Article 12, k(2) Position, San Antonio Water System, June 9, 2008.

115. Chad Norris et al., Edwards Aquifer Recovery Implementation Program, Senate Bill 3, Article 12, k(2) Position, Texas Parks and Wildlife Department, June 9, 2008.

116. See Moore and Votteler, Draft Habitat Conservation Plan, 26–28.

CHAPTER 6

The Importance of Freshwater Inflows to Texas Estuaries

Paul Montagna, Ben Vaughan, and George Ward

"I have always heard, Sancho, that doing good to base fellows is like throwing water into the sea." — Miguel De Cervantes, *Don Quixote*

*E*stuaries are coastal watercourses. They are physiographic indentations on the coastline, inundated by the postglacial rise in sea level, into which fresh water from the terrestrial environment drains. The great variety in geomorphological forms of estuaries arises from combinations of factors, including the character of the coast; the mechanism that created the coastal indentation (e.g., wave attack, scour by river discharge, faulting, volcanism); the intensity and trajectory of tidal currents; regional subsidence; the source, magnitude, and variability of freshwater inflow; and the transport of sediment. In most estuaries, freshwater inflow is delivered predominantly by organized terrestrial drainageways such as streams and rivers.

The essential feature of an estuary is that it is a transitional environment between freshwater and marine systems. It therefore is subject to processes of both terrestrial and oceanic origin and is a zone of concentrated gradients, notably in organics and nutrients (high in the terrestrial environment, low in the marine) and salinity (zero in fresh water and about 35 parts per thousand in the ocean). The resulting habitats make estuaries one of the most productive environments on earth.[1] Because estuaries are sheltered, productive systems that afford access to the sea, they are attractive to a variety of organisms that can tolerate or even exploit their transitional character. One such organism is *Homo sapiens*, which has historically colonized land adjacent to estuaries and upstream on their feeding rivers.

Numerous rivers and streams drain into the Texas coastline (Figure 6.1), and a variety of estuary types are associated with these basins. Most important for the present discussion are the major bays, identified in the figure. These are actually bay complexes, typically with two morphological components: a primary bay behind the barrier island connected to the sea and one or more secondary or tertiary bays into which streams and rivers flow.

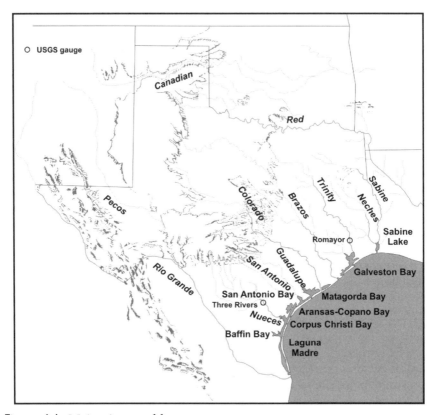

Figure 6.1 *Major rivers and bays*

Freshwater inflow is one of the defining characteristics of an estuary.[2] The fact that fresh water plays numerous roles in the processes operating in estuaries indicates that quantitative relations exist between the level of inflow and other aspects of estuary function, especially its productivity. Yet the extraction of such relationships and the definition of freshwater inflow needs of an estuary remain an unresolved challenge in many regions.[3] This chapter represents a brief survey that summarizes current understanding of the importance and ecological value of freshwater inflow to Texas estuaries and (with loose allusion to the quest of Don Quixote) elucidates the reasons it has been so difficult to quantify a link between inflow and measures of estuary health such as productivity.

TEXAS COASTAL HYDROLOGY: THE DUBBING

In Texas, rainfall—and hence runoff and riverflow—originates primarily from thunderstorms associated with seasonal disturbances. The intensity of these events is determined by the strength and latitudinal position of the westerly winds, their changes when they traverse the Rocky Mountain massif, the amplitude of air pressure disturbances carried by the westerlies, and the onshore influx of moist, unstable air from the Gulf of Mexico.[4] These facts have several implications. First, the time record of streamflow in a Texas river is a low base flow on which are superposed storm hydrographs, characteristic time signals with a relatively swift rise in flow followed by a prolonged decline. Figure 6.2 shows examples of one year of daily flow measurements for the Trinity and Nueces Rivers (the two gauge locations are indicated with circles in Figure 6.1).

Second, geographic variation occurs in the relative interaction of the meteorological controls on rainfall. The resulting gradient in rainfall (and thus riverflow) is from the arid Southwest to the humid Northeast, ranging on average from less than 25 centimeters (10 inches) per year in the Trans-Pecos to nearly 150 centimeters (60 inches) at the Louisiana border. Third, the relative interaction of the meteorological controls also has a seasonal variation that results in seasonality in rainfall and riverflow. Much of the central region of Texas exhibits spring and fall maximums in rainfall, the former increasing and the latter decreasing with latitude (i.e., to the north), while the western portion of the state has a summer maximum and the eastern a winter maximum.

During the high-rainfall seasons, the storm hydrographs in the streamflow record occur more frequently and are larger in magnitude. The clustering of larger, more densely distributed storm hydrographs is referred to here as the seasonal freshet. Conversely, during the low-flow seasons, storm hydrographs become smaller and sparser in time and space. There is an analogous geographic variation. In the high-rainfall, high-streamflow northeast region of the state, the frequency and intensity of storm hydrographs are greatest. With distance to the west and south, these storm events become sparser in time and space, and generally of smaller magnitude.

In addition to its annual seasonality, which also is highly variable from year to year, rainfall in Texas exhibits long period time variations, which can be characterized as prolonged alterations in the frequency, intensity, and timing of storm hydrographs. Under wet conditions, the rivers become prone to flooding. Under drought conditions, the paucity of storm runoff events leads to diminished and unreliable riverflows. The need to average these extremes in riverflows to service human development has led to extensive reservoir construction in Texas for both flood control and water supply.

A reservoir has two effects on the riverflow downstream from its dam. First, it induces a net loss of water to the river channel. The loss is due partly to enhanced evaporation (mainly because of the prolonged detention of water in the reservoir compared with the residence in the stream channel) and partly to diversions. Reservoirs offer the possibility of providing a water volume that can be dependably drawn on, which encourages water diversion and consumption. Indeed, this is the purpose of a water supply reservoir. Second, the reservoir induces a change in the time signal of streamflow, because storm hydrographs will be accumulated and detained in a reservoir. The natural flow in the river channel, subject to seaward slope of the channel bed and the mechanics of open-channel flow, is replaced by a constrained flow determined by the stage of the reservoir at the dam and the hydraulic properties of the spillway structure. If the storage capacity in the reservoir is large enough, the storm hydrograph may be entirely absorbed.

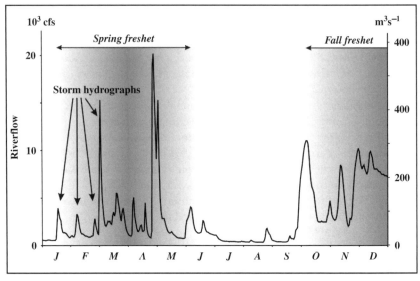

(a) Trinity River at Romayor

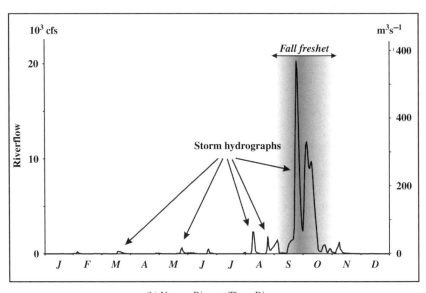

(b) Nueces River at Three Rivers

Figure 6.2 *Daily flows in 1964 at gauges on Trinity and Nueces Rivers, showing storm hydrographs and seasonal freshets*

Notes: cfs = cubic feet per second; m³ s⁻¹ = cubic meters per second.

The flows in the streams and rivers reaching the Texas coast are composites of the variety of hydroclimatologies and the operation of flow control structures (mainly dams) in their watersheds. Most river basins span a range of precipitation climates, with a generally diminishing runoff and a shift in rainfall seasonality with distance inland. The decreasing runoff across the state with distance to the south and west translates to a pronounced decline in flows from north to south along the Texas coast. This is exemplified by the data in Table 6.1, showing the average inflows into the principal bays during 1941–1999 and 1950–1956, the latter of which was one of the severest drought periods of the twentieth century for this region. From San Antonio Bay northward, it is the drought of record. The Lower Laguna Madre receives substantial flood flows from the Rio Grande diverted through the North Floodway and the Raymondville Drain. These are included in the inflows of Table 6.1, but the Rio Grande basin is not included in the watershed area.

TEXAS ESTUARIES: ALDONZA

The major estuaries of Texas are broad, shallow bays, as seen in Figure 6.1; their physical dimensions are summarized in Table 6.2. Their exchange with the Gulf of Mexico is restricted to inlets through barrier islands (which for all but Sabine Lake are extensive sandbars). With distance southward on the Texas coast, these inlets diminish in cross section, so the volume of tidal exchange diminishes as well. Combined with increasing aridity, decreasing rainfall, and diminishing inflow, the net effect is that these bays become increasingly saline southward along the coast, with salinities in the Laguna Madre and Corpus Christi Bay routinely exceeding that of seawater.

Humans have greatly modified the morphology of these estuaries, hardening shorelines by placement of bulkheads and riprap, making the landscape impervious through urbanization of the immediate watersheds, and dredging numerous channels for the benefit of ship traffic, notably the deep-draft ship channels in all the estuaries except San Antonio and Aransas-Copano Bays. The shallower Gulf Intracoastal Waterway, which runs from bay to bay, parallels the entire Texas

Table 6.1 *Inflows to the major bays*

Estuary	Watershed area (10^3 sq km)	Average inflows (10^6 m^3/yr)			
		1941–1999		1950–1956	
		Gauged	Ungauged	Gauged	Ungauged
Sabine Lake	54.1	14,290	2,933	10,900	1,841
Galveston Bay	63.5	9,922	3,516	4,308	675
Matagorda Bay[a]	130.3	2,926	1,244	907	335
San Antonio Bay	28.1	2,479	466	798	171
Aransas-Copano Bays	7.2	167	385	33	117
Corpus Christi Bay	45.6	644	102	342	91
Laguna Madre[b]	29.2	391	313	312	155

Notes: m^3/yr = cubic meters per year; sq km = square kilometers.
Source: Texas Water Development Board, http://midgewater.twdb.state.tx.us/bays_estuaries/hydrologypage.html (accessed September 13, 2010).
[a] Excluding East Matagorda Bay.
[b] Combined Upper Laguna, Lower Laguna, and Baffin Bay.

Table 6.2 *Physical dimensions of the major bay-estuary systems*

Estuary	Area (km²)	Volume (10⁶ m³)	Tidal prism[a] (10⁶ m³)	Residence time (days)
Sabine Lake	243	607	41	9
Galveston Bay	1,360	2,599	210	40
Matagorda Bay	948	2,217	144	81
San Antonio Bay	531	696	81	38
Aransas-Copano Bays	539	866	82	360
Corpus Christi Bay	497	1,188	65	356
Laguna Madre	891	511	36	—

Notes: km^2 = square kilometers; m^3 = cubic meters.
Sources: Hydrographic data from the National Ocean Service; residence time from R.S. Solis and G.L. Powell, Hydrography, Mixing Characteristics, and Residence Times of Gulf of Mexico Estuaries, in *Biogeochemistry of Gulf of Mexico Estuaries*, ed. T.S. Bianchi, J.R. Pennock, and R.R. Twilley, 55 (New York: John Wiley & Sons, 1999).
[a] The volume of water entering the estuary on the flooding tide.

coastline. Inlets between the Gulf of Mexico and the bays have been created, plugged, deepened, widened, narrowed, and partially shoaled. Jetties on the seaward termini of inlets have greatly modified the drift of sand along the beachfront and the interception of drifting sand and its transport through the tidal inlets. Within the bays, placements of dikes, levees, and dredge spoil have resulted in major alterations of internal circulation patterns.

The processes operating within an estuary are conveniently classified into four categories: geology, including the movement of sediments into and out of estuarine boundaries, subsidence, and tectonic movements; hydrography, encompassing the movement of water within the estuary; chemistry, involving the concentrations and kinetics of substances carried in solution or suspension in the estuary waters; and biology, as it relates to the totality of organisms living within and on the periphery of the estuary. This sequence of physical and biological features also generally follows the sequences of cause-and-effect relationships. Figure 6.3 is a highly simplified schematic of cause-and-effect connections among hydrography, chemistry, and biology in the Texas estuaries. Nothing is implied about the nature of the relationships or their time and space variations within the estuary. The figure does not include geologic processes, because they operate on much longer timescales than the processes shown and are manifested as the boundary controls of physiography and bathymetry (i.e., distribution of water depths), which are included in morphology.

The hydrography of the Texas estuaries can be summarized as follows. Tidal variation is small but nonnegligible. The maximum tide range in the Gulf of Mexico is about 1 meter. This tide is attenuated substantially as it passes through the inlets, especially in the higher frequencies (such as the twice-daily high- and low-tide components). The water levels in Texas bays are responsive to meteorological forcing, a consequence of dynamic alternation between the northwesterly winds that drive continental air masses and the southeasterlies of the trade winds. Strong winds interact with the shallow physiography of the bays, causing setup, a slope in the water surface from the windward to the leeward side of

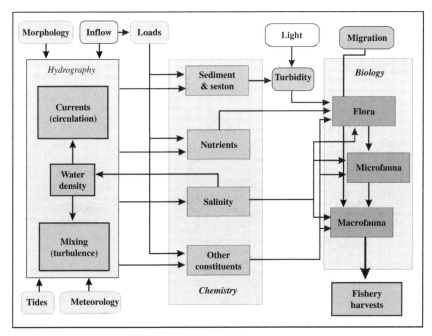

Figure 6.3 *Simplified schematic of causal controls on biology in an estuary*

the bay. A semiannual rise and fall in coastal water levels also occurs, with maxima in spring and fall and minima in summer and winter. These seasonal "tides" are responsible for substantial exchange between the estuaries and the Gulf. In addition to these volume exchanges, the estuary waters are diluted and occasionally replaced by high-inflow events. Salinity intrusion from the Gulf of Mexico into bays is effected by tidal and meteorological exchanges with the sea. More important, salinity intrusion also results from currents caused by denser salt water moving from the sea into the estuary in zones of deeper water, displacing estuary water seaward. These circulations are driven by the seaward gradient in water density, because of the density difference between fresh and salt water.

The combination of hydrographic processes makes data interpretation difficult. For example, salinity, for which suboceanic values ultimately are due to freshwater dilution, does not exhibit a clear association with the level of inflow because of the variable time response to other processes such as salinity intrusion or evaporation. For other parameters that are subject to kinetic reactions as well as transport and dilution, the variation is even more complex.

As in any aquatic system, a variety of organisms live in and about the estuary waters. From a "who eats whom" viewpoint, the estuary ecosystem has three main trophic subsystems: producers, consumers, and decomposers (see Figure 6.4).

Producers can generate biomass and energy from sunlight and carbon; these are the plants, or autotrophs (represented by "flora" in Figure 6.3). Consumers, including animals and most bacteria, are incapable of generating their own energy or biomass and must eat producers or other consumers; these organisms are

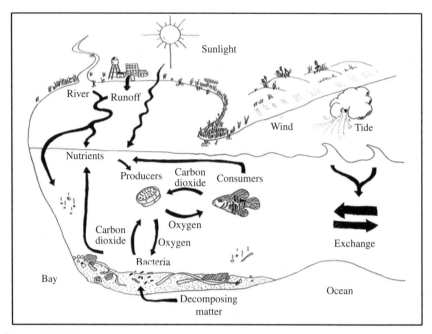

Figure 6.4 *Estuary ecosystem*

Source: P.A. Montagna, J. Li, and G.T. Street, *A Conceptual Ecosystem Model of the Corpus Christi Bay National Estuary Program Study Area*, Publication CCBNEP-08 (Austin, TX: Texas Natural Resource Conservation Commission, 1996), http://www.cbbep.org/publications/virtuallibrary/ccbnep08.pdf.

heterotrophs. Most organisms, producers and consumers alike, also require oxygen for respiration and thus are aerobic. Those that do not require oxygen are anaerobic. Decomposers, which include microscopic bacteria, are consumers that respire by decomposing detritus, or dead organic matter. They liberate carbon dioxide from the organic matter in the process and hence are also called mineralizers, and they can be aerobic or anaerobic. They "close the loop" by breaking down complex organic molecules created by autotrophs and consumers into simpler inorganic compounds.

The larger consumers (separately identified in Figure 6.3 as "macrofauna") include the finfish and shellfish, which are the most conspicuous expression of estuarine productivity. Estuary macrofauna have two characteristics that distinguish them from those in other aquatic ecosystems: they must be capable of functioning in the highly variable environment of the estuary, and many of them migrate between the ocean and the estuary at various stages of their life cycle.

The first means that the transitional nature of the estuarine environment excludes purely freshwater or marine species. The primary parameter that measures this relative hospitality of the estuary is salinity, which can be thought of as a measure of the proportion of seawater mixed with fresh water. It is the nature of estuaries that salinity ranges spatially and temporally from zero (fresh water) to at least oceanic (seawater). The mixing zone from fresh to marine, the main gradient of salinity in an estuary, can move horizontally over very small timescales with daily tides or longer timescales with seasonal tides, and it can expand or contract within the estuary, depending on conditions. The volume of fresh water diluting seawater also varies with time, both seasonally and from year to year. This means that the range of variation of salinity is high, leading to an appreciable difference between fresh water and seawater over short time periods and large spatial scales. Thus, estuarine organisms that are immobile or move more slowly than the mixing zone must have much wider salinity tolerance limits than do purely fresh or marine species. Moreover, organisms historically abundant in an estuary must have adapted to historical salinity ranges, especially at regional scales. Biologically, the salinity gradient is the main factor controlling an estuarine community and population structure.

Most of the macrofauna populations in an estuary are transient, and most of these migrate into the estuary in their early life stages, seeking shelter and food to allow them to grow into adults, whereupon they migrate back to the ocean. The National Marine Fisheries Service estimates that 75% of the nation's commercial fish and shellfish depend on estuaries at some stage in their life cycle, and that 98% of the commercial fish and shellfish harvested in the Gulf of Mexico are dependent on estuaries.[5] Biologically, the importance of the Texas estuaries extends far beyond their geographic bounds.

ECOLOGY OF TEXAS ESTUARIES: TO EACH HIS DULCINEA

Estuary ecosystems can be perceived in several ways. Trophic function and food web analysis depicts the interdependencies of estuarine populations. Estuary function can be described by the processing of organic and inorganic materials through various subcomponents of the ecosystem. The main components of estuarine ecosystems are the system itself (with many subcomponents), its input sources, and output sinks (Figures 6.3 and 6.4). Terrestrial runoff, including riverflow, transports nutrients for primary producers and detritus for consumers and mineralizers. It also can be useful to analyze the transfer of energy through the estuary, from external sources (primarily the sun) to its repackaging into organic material, which is transferred through the ecosystem and exchanged with the coastal ocean via currents.

From the standpoint of projecting impacts on the estuary environment, a useful concept is that of habitat. A habitat is the suite of biological, physical, and chemical attributes that sustains a species or community of organisms. A community is composed of different species coexisting in a habitat. In some cases, organisms define the habitat, such as marsh plants that create marsh habitat. In other cases, physical elements define the habitat, such as the water column for open bay habitats. The complex physiography of the Texas coast creates diverse estuarine habitats, and the availability of these habitats may be essential for certain species. The link between inflow and the ecological structure and function of estuarine habitats is through the physicochemical factors that are changed when the inflow regime is altered, such as the salinity gradient, nutrient concentrations, and sediment loading (Figure 6.3). The structure and function of the habitats are not determined by freshwater inflow in and of itself. Inflow drives the physicochemical conditions of the estuary, and it is the estuarine condition that defines the habitats that control biological response. The diversity of estuarine habitats is the main reason that there is high biological productivity in estuaries. In the major bay ecosystems of Texas, typical estuarine habitats include river, salt marsh, algal mat, seagrass bed, water column, open bay bottom, oyster reef, beach, coastal ocean, and others, as depicted in Figure 6.5. Some habitats are physical in nature, but others, such as reefs and wetlands, are created by biological species themselves. The habitats also are often arranged along the geomorphological gradient in the estuary from the river to secondary bay to primary bay.

Although each habitat may appear distinct, many interconnections exist among habitats. Water currents, waves, and tides transport organic matter, energy, and animals from one habitat to another, and many types of animals can move among many different habitats. Different life stages of many species will move among the habitats during the different phases of their life cycles. The interactions and connectivity among habitats in the river-to-sea gradient are partly responsible for estuaries' high productivity, as well as the ecological services that benefit humankind.

All Texas estuaries have a common structure similar to that illustrated in Figure 6.5. Exchange with the Gulf of Mexico occurs through a pass in the barrier island. Beach habitat faces the ocean or barrier island. The Gulf is connected to a primary bay with a bottom that is predominantly a muddy habitat, with patchy areas of sandy bottom or oyster reefs. Oyster reef habitats frequently occur in

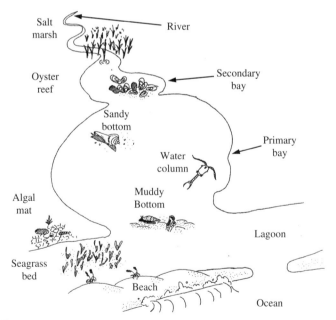

Figure 6.5 *Estuarine geomorphological components and habitats*

Source: Montagna et al., *Conceptual Ecosystem Model.*

secondary bays or near the junctions of primary and secondary bays. Rivers empty into the secondary bays, and sometimes tertiary bays or lakes are also associated with the rivers. Marshes line the river sources of tertiary and secondary bays. Lagoonal sounds run parallel to the barrier islands and perpendicular to primary bays. From the Laguna Madre to Matagorda Bay, the primary bays are connected by the lagoonal sounds, which therefore are important for exchange of materials and recruitment between systems. In the southern bays, algal mats develop on broad supratidal flats.

Seagrasses grow in areas characterized by oceanic levels of salinity and shallow depths. Seagrass habitats support a very diverse and productive food web by providing a source of carbon for the food web and a place for fish and invertebrates to hide from predators. The high amount of biomass from these plants leads to high rates of gross primary productivity and net community productivity. Seagrass is also a substrate for epiphytic algae (microalgae that grow on seagrass blades) and animals such as crustaceans and polychaete worms. Beds of seagrass play an important role as nursery grounds for larval fish and invertebrates. They also serve as buffers against storms and can help filter contaminants from the water.

Marsh habitats are located in intertidal regions of the bay and are often expansive near freshwater sources, secondary bays, or fringing wetlands within primary bays. They are dominated by smooth cordgrass (*Spartina alterniflora*). Marshes are important because they trap sediments, store water, buffer the coast during storms, and filter wastes from water. Most important, marshes provide a nursery environment for the young of many species, offering food sources and shelter from predation and harsh conditions. For many years, marshes were thought to function as nutrient exporters to the larger estuarine system and therefore to be an important control on primary production. However, quantitative field studies in several major estuaries, including those in Texas, indicate a more equivocal role of marshes in the nutrient budget, sometimes representing a source, but often serving as a sink that traps nutrients and enhances water quality.

Pelagic habitats are the water columns that fill all estuaries. The water column can become quite turbid as sediment is resuspended by wind waves or by human activities that disrupt the bed, such as dredging or trawling. Because fresh water mixes with salt water in the bays, the water column typically has an extensive zone of brackish salinities (10 to 25 parts per thousand). When evaporation exceeds freshwater inflow and flushing by the ocean, salinities can exceed that of the ocean. The water column is usually well oxygenated, because of the broad, shallow, wind-mixed character of these bays. The food web consists of phytoplankton being eaten by zooplankton, which in turn are eaten by fish. This grazing food web dominates the water column, in contrast to seagrass and marsh habitats, which are dominated by the detrital food web. Primary production by phytoplankton in estuarine water can be relatively high.

Mud and sand are common substrates beneath the pelagic zones of all Texas bays. Sand can support larger animals that might sink in soft mud. Sandy bottoms often are accompanied by stronger currents and higher water transparency than are found in muddy water habitats. Muddy bottoms are by far the most common benthic habitat in Texas bays. The dominant species are small benthic invertebrates, such as clams, polychaete worms, and crustaceans. These animals are important food sources for edible shrimp (*Penaeus*), which support a large commercial fishery. Shrimp are eaten by a variety of fish, such as catfish (*Arius felis*), red drum (*Sciaenops ocellatus*), and flounder (*Paralichthys lethostigma*). Small clams are a primary food source for black drum (*Pogonias cromis*). In shallow pelagic areas where light penetrates through the water column, benthic algae can contribute to community production.

In Texas bays, the most common hard bottom habitat is created by oysters (*Crassostrea virginica*). Oysters flourish in shallow water of intermediate salinity, and their shells form extensive reefs in secondary bays with high freshwater inflow. These reefs have two dramatic habitat effects. First, they provide natural hard bottom habitat for encrusting fauna. Second, the physical structure of the reef acts as a barrier to water flow, which can cause organic matter to settle out of the water and onto the reef, where it can fuel a detrital-based food web. Many species in oyster reefs are filter feeders, including the oyster itself and animals that encrust oyster shells. These filter feeders are the natural vacuum cleaners of estuaries and help maintain the water quality of bays. With such a high biomass and diversity of food sources, oyster reefs support a large and diverse food web.

A suite of chemical parameters also affects the quality of the habitat and in many cases enables its existence. Among the defining parameters is salinity. Although estuarine organisms are capable of withstanding a wider range of salinity than their freshwater or marine kin, they are affected by salinity. Most of them do have limits on salinity tolerance and optimal salinity ranges for growth, development, or reproduction. Salinity also can influence foraging or reproductive behavior as organisms seek suitable habitats. The two most important material-conversion processes affected by salinity are primary production and decomposition. Most plants have optimal salinity ranges for photosynthesis, and salinity is usually an inverse indicator of the availability of land-derived nutrients, often constraining primary production.

QUANTIFICATION OF THE IMPACT OF FRESHWATER INFLOWS: THE IMPOSSIBLE DREAM

Several facts about estuary hydrography and ecology emerge from the above summaries. The estuary presents a highly variable environment. This variation is a source of physiological stress that limits or even excludes many organisms, but it creates a large variety of niches that allow others to thrive. The major physical controls on the estuary are the ultimate source of this variability. Freshwater inflow is an important control. The diagram of Figure 6.3 allows a simplified mapping of the causal relations involving biology, in which freshwater inflow plays a dominant role. These include the following:

- conveyance of terrestrial-origin nutrients;
- dilution of marine salinity (modulated by estuary transport processes);

- conveyance of terrestrial-origin sediments, especially fine particulates;
- inundation and flushing of important zones, such as marshes, by flood events; and
- source of time variation in estuary properties.

Freshwater inflow is, however, just one of several controls (also shown in Figure 6.3), and the others can have equal or even greater influence on the estuary. The potential human impacts on freshwater inflow to the estuaries are not trivial. This is exemplified by Table 6.3, which summarizes the long-term proportionate change due to human activities in average inflow to the principal rivers entering the major Texas bays. These results have been obtained from the Texas Water Availability Model (WAM), operated to establish "naturalized" and "current (ca. 2000)" flows.[6] Naturalized flows are those resulting when the effects of human "plumbing" are removed. These include reservoir operations and the associated diversions, as well as agricultural water use and return flows (which can represent a net water source to a river when groundwater is the primary water supply or when water is transferred into the basin from another watershed). The reduction in inflow due to these hydrologic impacts can be substantial, the largest being 44% for the Colorado River, 95% of whose watershed is regulated by reservoirs. Human effects during extreme drought conditions are exemplified by the 1950s drought period, also presented in the table; these effects are even more severe in the short term during drought periods than in their long-term average impact, especially in the lower bays. This is because the atmospheric and hydrologic processes during a drought are correlated and reinforcing, and this is compounded by human impacts.[7]

Quantitative hydrologic accounting, such as the WAM model, is capable of determining the modifications to freshwater inflow under various past or future scenarios of human activity. The problem, however, lies in translating this quantitative impact on inflow to a quantitative impact on the estuary ecology Approaches in the literature devolve into five general strategies, presented here in order of increasing complexity and sophistication:

1. Determine the proportional human impacts on inflow, and then assume that the same relative impact will translate to a biological response.
2. Determine the hydrographic and chemical effects of changing inflow, and then assume these will directly translate into biological impacts.
3. Establish a statistical relationship between inflow and one or several measures of the biological community.
4. Determine the effects of changing inflow on salinity (and perhaps other chemical constituents), then use tolerance ranges of selected species to define habitats, and predict changes in size (and perhaps location) of habitats.
5. Formulate a detailed process model of the ecosystem, of which inflow is one component.

An example of approach 1 is the percentage-of-flow approach, in which freshwater inflow alterations are limited by a prespecified proportion of some key hydrologic statistics, such as long-term mean inflow. In southwest Florida, for example, total inflow reduction to an estuary is limited to 10%.[8] The most

Table 6.3 *WAM-projected change in inflow (%) relative to naturalized major river inflows to major bays*

Bay				
Sabine Lake	Galveston	Matagorda	San Antonio	Corpus Christi
River inflows				
Sabine and Neches	Trinity	Colorado	Guadalupe and San Antonio	Nueces and San Antonio
Current development impact on inflows				
1940–1996 –7	1940–1996 –24	1940–1996 –44	1934–1989 –8	1948–1996 –29
1949–1956 –8	1949–1956 –32	1949–1956 –77	1949–1956 –11	1949–1956 –43

Sources: WAM output for the Colorado: TRC/Brandes unpublished data 2006; WAM output for all other rivers: Texas Commission on Environmental Quality unpublished data 2009.

common parameter addressed in approach 2 is salinity, examples of whose use as a surrogate for inflow are San Francisco Bay,[9] the Suwannee estuary,[10] and Matagorda Bay.[11] This approach requires substantial historical data on salinity, whether its relation to inflow is established by statistical association or hydrodynamic modeling. The relation to biology, however, often is frustrated by the euryhaline (ability to tolerate a wide range of salinities) or mobile nature of many estuarine organisms. The Texas methodology exemplifies approach 3 and is discussed below.[12] Approach 4 is in common use in Florida, such as in the Caloosahatchee Estuary, where the goal was to create salinity zones to protect seagrass habitats.[13] The demands of approach 5, both in theoretical support and data availability, make it daunting. However, Kremer and Nixon provide an example implementation for Narragansett Bay,[14] Montagna and Li for the Texas coast,[15] and Chan et al. for Western Australia.[16] Hydrodynamic and transport modeling frequently play a role in literature examples of approaches 2, 4, and 5, but careful evaluation of field data can suffice if the database is adequate.

Two theoretical elements are needed to quantify freshwater inflow effects on an estuary. First, an indicator of the ecological state of the estuary must be formulated quantitatively through some objective metric. It is mandatory that this metric be measurable, or derivable from measurable variables. Second, functional relationships must be established between freshwater inflow and the ecological metric. Of the literature approaches summarized above, 2 through 5 more or less incorporate these elements. Approach 1 is a management artifice resorted to in lieu of quantitative knowledge of the estuary and relies only on the data record of inflows.

With respect to the first element, probably the most fundamental quantification of the estuary ecology is the size of a population of a species in the system—that is, its abundance, typically expressed as either an areal abundance or volumetric density of individuals. Abundance of a species is a strong function of both position in the estuary and time, with shorter-term fluctuations superimposed on a seasonal variation in response to climatology and hydrology, as well as migration for many of the motile animal species. The metric may be abundance simply of a single species (or perhaps several) that is of social or economic importance or that is useful as a sentinel. If abundance data for a large number of species are available, then the data can be combined into parameters of structure, such as diversity or trophic ratios. Abundance information can be refined by space–time dependence and augmented with data on the age, weight, condition, tissue chemistry, and food of individuals of the selected species. The important point is that the metric is either directly observed counts of organisms, perhaps augmented by physical or chemical measurements on specimens, or parameters derived from such data by conversion and numerical combination.

As a quantitative concept, abundance is very important but not without limits. Abundance of indicator species can point to specific zones of influence that alter with changes in flows. For example, insect larvae are highly abundant during floods in freshwater segments but never present in marine waters. In contrast, brittle stars are present only when marine conditions persist over long periods of time. In this way, indicator species can be useful to identify specific salinity zones.[17] However, as pointed out earlier, these zones are highly variable in space and time.

The combined abundances of the different species define the biodiversity of the area or zone. Diversity is an important component of the stability and long-term sustainability of an ecosystem, and thus diversity metrics have long been important ecological tools. In fact, diversity indices are often misused. A simple diversity metric by itself is not very useful, because it contains no information about the indicator species or any species of interest. Moreover, in an estuary, diversity by itself can be highly misleading, because many major species occur transiently in large numbers, thereby depressing the numerical value of diversity. The value of biodiversity as an indicator is the multivariate data available for all the species, not the univariate reductionist index of diversity.

A qualitative concept that is often promoted as a metric in the context of ecosystem impacts is productivity. There are many dimensions to productivity and many interpretations. Total ecosystem productivity is a function of primary and secondary production, neither of which is conveniently measurable. Biomass alone sometimes is used as a measure of productivity,[18] in which case it is really akin to abundance. One problem with productivity as a metric is that most people may not care about primary producers, but rather be interested only in some consumers, such as certain species of fish. More generally, the value of productivity per se is often less important than which ecosystem component is doing the producing. For example, early studies on the effects of acid rain found that productivity did not change, but acidic lakes were producing carp and not trout.[19]

Other conceptual indicators of ecological state exist as well, a large class of which have qualitative accord with desirable attributes of an ecosystem and therefore rhetorical power in sociopolitical and environmental management contexts. A frequently cited example is ecosystem health, or as the Texas legislature put it, a "sound ecological environment." Other examples are sustainability, integrity, and resilience. These last three can be related as follows. Ecological integrity can be defined as the condition that exists when measurable physical, chemical, and biological parameters (or metrics formed from these parameters) fall within an acceptable range. Sustainability is ensured when ecological integrity is maintained over time. An estuary is resilient if the parameters of interest return to their acceptable ranges after natural or human-caused disturbances. The crux of all these definitions is our ability to quantify an "acceptable range." A surrogate, pending more research, might be a condition that densities of animals and plants are not significantly different from historical patterns of abundance or from recently observed patterns. But the lack of precision in the definition of ecosystem health (to say nothing of logical circularity) continues to frustrate ecology-based management.

Were the problem of definition solved, that of insufficient data would still remain. All of the above candidate metrics share a common difficulty: they require adequate field data for their evaluation. As a rule, field biological data are collected by manual methods (e.g., sediment samples, water samples, entrapment gear), require painstaking expert knowledge to identify organisms, and therefore are labor-intensive. This causes such data to be discrete in space and time, sparse, and biased toward daylight and fair weather, as well as to exhibit substantial gaps in long-term coverage. The populations observed, furthermore, are patchy in space and time, so the measurements, which are instantaneous point samples of a highly variable function, have considerable uncertainty.

The second necessary element in quantifying freshwater inflow effects is the functional relationship between freshwater inflow and the selected ecological measures. Figure 6.3 indicates a number of potential causal pathways among controls on the estuary, only one of which is inflow, and the response of hydrographic, chemical, or biological components. Our ability to evaluate any of these pathways and assess their relative importance to various biological organisms requires a quantitative expression of any underlying causal relationships. Ideally, such a relationship would be expressed as a deterministic principle. The state of the sciences is that such determinism is limited to the most elementary processes of hydrography, chemistry, and organismal physiology. Otherwise, these relationships must be sought empirically, by quantifying the associations among measurements of the different variables involved. In most cases, as one progresses from the upper left corner of Figure 6.3 to the lower right, the analysis becomes increasingly empirical and statistical.

The dominance of empiricism has three immediate and profound consequences. First, the validity of any such relationship is circumscribed by the available data. The success in delineating the dependencies of the estuary ecosystem on inflow (or, for that matter, any other external factor) is directly dependent on the quality of the sampling design (i.e., monitoring strategy) and the extent and quality of measurements made. Among other things, these measurements must be of adequate density in space and time to resolve these variations, and they must be sustained sufficiently long to reflect the natural temporal variability in all of the variables. Second, the sampling design must contain sufficient controls and treatments, and avoid confounding factors and pseudoreplication; otherwise, it will be impossible to interpret the results using logic, and we will fall into the "hypothesis falsification" trap. The fact that the magnitude and variation of most of the factors (as shown in Figure 6.3) are out of our control compounds the difficulty of sampling design. Practically, this entails data collection being sustained over a long enough period and extensive enough areas to encounter the requisite ranges and combinations of variables. The limited interpretation of poor experimental designs is the subject of an enormous literature; see, for example, Underwood[20] and Stephens et al.[21] Third, because our ability to extract an underlying relationship is undermined by variance in the data, the high variance implicit in the estuarine environment imposes a formidable obstacle. Additional sources of variance include imprecision in the measurements and random errors. Two other challenges are variables not being recognized or taken into account in the evaluations of statistical association; and inadequacies in the assumed mathematical form of the empirical relationship between inflow and response variables.

The state of Texas, through two of its primary resource agencies, the Texas Water Development Board (TWDB) and Texas Parks and Wildlife Department (TPWD), has made a valiant attempt to determine the inflows "adequate to maintain an ecologically sound environment." This state methodology is documented in Longley[22] and Powell et al.[23] and has been applied to each of the major bays of Texas. The core of the methodology is to determine linear statistical regressions of measures of the abundance of several key organisms—black drum, flounder, blue crab, white shrimp, brown shrimp, and oyster—on freshwater inflow; therefore, it is a variation of approach 3. In Longley and early applications of the methodology, the measure of abundance was taken to be annual commercial harvest of the species.[24] More recent applications have abandoned harvest in favor of fishery–independent catch data of the TPWD. A novelty of the state approach is to apply these relations to the determination of an optimum level and seasonal variation of inflows.[25] Whether optimization of fisheries species is an appropriate determination of freshwater inflow needs of an estuary is rendered questionable for two reasons: the model ignores economic and social drivers, and ecosystems do not optimize productivity.

In addition, several technical problems are inherent in the optimization approach. First, the optimization model produces a pattern of monthly inflows, which, though deemed optimal, are not found in the historical record, even to a rough approximation. Second, the lack of normalization of the commercial harvest data for variations in effort means that changes in external variables (e.g., the price of fuel) may be conflated with the impact of changes in freshwater inflows. Finally, the optimization solution must be bounded within realizable values of the variables by imposing constraints. The state methodology has two important constraints on its solutions: the monthly inflows have a limited range (between the first decile and the median), and there cannot be a surfeit of one species. These two constraints cause the methodology to overlook the innate variability of inflows and the impact this has on the quantity and diversity of estuarine production.

Notwithstanding the above, the state continues to make inflow policy, and Senate Bill 3, passed in 2007, requires all Texas bays to have inflow criteria, which will be developed by bay and basin stakeholder committees. It is incumbent on the scientific community to inform this process.

Given the state of scientific knowledge, the best available candidates for measuring the effects of freshwater inflows in estuaries are physicochemical changes in hydrography (including nutrients, dissolved oxygen, and salinity), the areal extent of estuarine habitats, and abundance of sessile organisms that respond to changes in salinity or habitat conditions. These might be expanded to include water quality, diversity of species, and metrics combined in an index that measures various aspects of biological community function and composition. Indicators of eutrophication (that is, nutrient overenrichment) such as hypoxia (oxygen deficiency) and algal blooms, concentrations of toxic compounds, and other water quality data routinely monitored by many organizations may also be useful in specific systems. For example, distributional shifts in the relative abundance of plant species often are associated with different environmental factors. A similar situation occurs for estuarine marsh communities; for example, saltmarsh bulrush is associated with lower-salinity conditions than is smooth cordgrass. Oyster reefs can also serve as environmental indicators, because oysters spend their entire adult life in the estuary as sessile (fixed) organisms. They are capable of surviving a fairly wide range of salinity conditions, but high salinities can limit oyster abundance in estuaries of the lower Texas coast. Oysters also are subject to predation by or disease from organisms (such as oyster drills and dermo, respectively) that thrive in high-salinity conditions, which significantly increases oyster mortality. More generally, species-based indicators can include abundance of key species, such as socially significant animals, fishery species, threatened or endangered species, sentinel benthic (bottom-dwelling) organisms, plant communities, or disease organisms. However, species-based indicators are valid only if the species are known to require narrow salinity ranges.

CONCLUSIONS: THE GOLDEN HELMET OF MAMBRINO

Current Texas policy is that environmental freshwater inflows to the state's bays and estuaries must be sufficient to protect environmental health. But relating flows to health is a doubly difficult task, because the processes affecting and operating within estuaries are complex (as Figures 6.3 and 6.5 illustrate), and the definition of health is system-specific. Definition of appropriate environmental flows to Texas bays and estuaries, then, will be a long-term endeavor. Two initial conditions must be met.

First, the relationship between freshwater inflows and the biological responses of bays and estuaries must be modeled empirically or deterministically (most likely a combination). This will require more extensive monitoring of environmental conditions in many systems, across both time and space, than has heretofore been done, as well as scrutiny of those data already collected. The estuarine environment is subject to a large range of natural variability (e.g., Figure 6.2), and data gaps exist about some responses, especially under extreme flood or drought conditions. Such a monitoring program will require substantially more funding than has been devoted to date. But without the results of such a model, the policies needed to define and maintain environmental health cannot be determined, nor can the costs of these measures be weighed against alternative, consumptive uses of water.

Second, the definition of environmental health requires a scientifically informed stakeholder process to identify valuable ecosystem characteristics that should be measured and conserved. Such a process has already begun under legislative mandate, albeit without the aid of extensive scientific data or understanding for some of the bay systems. This process will produce consensus about the metrics of health and their desired values, as well as much-needed insight into the baseline problem of bay and estuary health—namely, that the current state of each of the principal bays of Texas, from Sabine Lake to the Laguna Madre, is the standard for future regulatory action. Yet human activities already have reduced inflows by between 7 and 44% in the past half century, as compared with naturalized flows. Potential ecological baselines then include historical conditions (although data to delineate these may be limited), current conditions, or even some degradation from current conditions (supposing that our understanding of the bays' ecological function were to advance sufficiently as to be able to predict what this might be).

This chapter has identified measures and concepts that may help inform this process. Plant and oyster habitats serve as good environmental indicators, for example, because they are generally rooted in place, their extent is measurable, and they are relatively long-lived. Ecological integrity is the maintenance of measurable physical, chemical, and biological parameters (or metrics derived from them) within acceptable ranges as designated through a scientifically informed stakeholder process. As such, it combines social values, economic conditions, and ecological outcomes when used in conjunction with a detailed process model of the ecosystem.

Once the above goals have been achieved, only two problems will remain. Sufficient hydrologic control of the river basins must be established—and used to maintain environmental health, even in drought. And sufficient water must be purchased to be used for this purpose. Only then will we know whether Dulcinea is real or a figment of our imagination.

NOTES

1. E.P. Odum, *Fundamentals of Ecology*, 2nd ed. (Philadelphia: W.B. Saunders Co., 1959).

2. D.W. Pritchard, What Is an Estuary: Physical Viewpoint, in *Estuaries*, ed. G. Lauff, Pub. 83, 37–44 (Washington, DC: American Association for the Advancement of Science, 1967); G.H. Ward and C. Montague, Estuaries, in *Water Resources Handbook*, ed. L. Mays, vol. 12, 1–114 (New York: McGraw-Hill Book Company, 1996); K.R. Dyer, *Estuaries: A Physical Introduction* (Chichester, UK: John Wiley, 1997); P.A. Montagna, M. Alber, P. Doering, and M.S. Connor, Freshwater Inflow: Science, Policy, Management, *Estuaries* 25 (2002): 1243–45.

3. Montagna et al., Freshwater Inflow.

4. G.H. Ward, Texas Water at the Century's Turn: Perspectives, Reflections and a Comfort Bag, in *Water for Texas*, ed. J. Norwine, J. Giardino, and S. Krishnamurthy, 17–43. (College Station: Texas A&M University Press, 2005).

5. M.E. Patillo, T.E. Czapla, D.M. Nelson, and M.E. Monaco, *Distribution and Abundance of Fishes and Invertebrates in Gulf of Mexico Estuaries*, Vol. 2, *Species Life History Summaries*, ELMR Rep. No. 11 (Silver Spring, MD: NOAA/NOS Strategic Environmental Assessment Division, 1997).

6. R.A. Wurbs, Methods for Developing Naturalized Monthly Flows at Gaged and Ungaged Sites, *Journal of Hydrological Engineering* 11 (2006): 55–64.

7. Ward, Texas Water at the Century's Turn.

8. M. Flannery, E. Peebles, and R. Montgomery, A Percent-of-Flow Approach for Managing Reductions of Freshwater Inflows from Unimpounded Rivers to Southwest Florida Estuaries, *Estuaries* 25 (2002): 1318–32.

9. W. Kimmerer and J. Schubel, Managing Freshwater Flows into San Francisco Bay Using a Salinity Standard: Results of a Workshop, in *Changes in Fluxes in Estuaries: Implications from Science to Management*, ed. K. Dyer and R. Orth, 411–16 (Fredensborg, Denmark: Olsen and Olsen, 1994).

10. R.A. Mattson, A Resource-Based Framework for Establishing Freshwater Inflow Requirements for the Suwannee River Estuary, *Estuaries* 25 (2002): 1333–42.

11. Q. Martin, D. Mosier, J. Patek, and C. Gorham-Test, Freshwater Inflow Needs of the Matagorda Bay System (Austin, TX: Lower Colorado River Authority, 1997).

12. W. Longley, ed., *Freshwater Inflows to Texas Bays and Estuaries* (Austin, TX: Texas Water Development Board, 1994); G.L. Powell, J. Matsumoto, and E.A. Brock, Methods for Determining Minimum Freshwater Inflow Needs of Texas Bays and Estuaries, *Estuaries* 25 (2002): 1262–74.

13. P.H. Doering, R.H. Chamberlain, and D.E. Haunert, Using Submerged Aquatic Vegetation to Establish Minimum and Maximum Freshwater Inflows to the Caloosahatchee Estuary, Florida, *Estuaries* 25 (2002): 1343–54.

14. J. Kremer and S. Nixon, *A Coastal Marine Ecosystem: Simulation and Analysis* (New York: Springer-Verlag, 1978).

15. P.A. Montagna and J. Li, *Modeling and Monitoring Long-Term Change in Macrobenthos in Texas Estuaries, Final Report to the Texas Water Development Board, Marine Science Institute*, Technical Report No. TR/96-001 (Port Aransas, TX: University of Texas at Austin, 1996).

16. T.U. Chan, D.P. Hamilton, B.J. Robson, B.R. Hodges, and C. Dallimore, Impacts of Hydrological Changes on Phytoplankton Succession in the Swan River, Western Australia, *Estuaries* 25 (2002): 1406–15.

17. P.A. Montagna, and R.D. Kalke, The Effect of Freshwater Inflow on Meiofaunal and Macrofaunal Populations in the Guadalupe and Nueces Estuaries, Texas, *Estuaries* 15 (1992): 307–26; P.A. Montagna and R.D. Kalke, Ecology of Infaunal Mollusca in South Texas Estuaries, *American Malacological Bulletin* 11 (1995): 163–75; P.A. Montagna, R.D. Kalke, and C. Ritter, Effect of Restored Freshwater Inflow on Macrofauna and Meiofauna in Upper Rincon Bayou, Texas, USA, *Estuaries* 25 (2002): 1436–47.

18. K. Banse and S. Mosher, Adult Body Mass and Annual Production/Biomass Relationships of Field Populations, *Ecological Monographs* 50 (1980): 355–79.

19. D.W. Schindler, K.H. Mills, D.F. Malley, D.L. Findlay, J.A. Shearer, I.J. Davies, M.A. Turner, G.A. Linsey, and D.R. Cruikshank, Long-Term Ecosystem Stress: The Effects of Years of Experimental Acidification on a Small Lake, *Science* 228 (1985):1395–97.

20. A.J. Underwood, *Experiments in Ecology* (Cambridge, UK: Cambridge University Press, 1997).

21. P.A. Stephens, S.W. Buskirk, and C.M. del Rio, Inference in Ecology and Evolution, *Trends in Ecology and Evolution* 22 (2006): 192–97.

22. Longley, *Freshwater Inflows*.

23. Powell et al., Determining Minimum Freshwater Inflow Needs.

24. Longley, *Freshwater Inflows*.

25. Q.W. Martin, Estimating Freshwater Inflow Needs for Texas Estuaries by Mathematical Programming, *Water Resources Research* 23 (1987): 230–38; Y. Tung, Y. Bao, L. Mays, and G. Ward, Optimization of Freshwater Inflow to Estuaries, *Journal of Water Resources Planning & Management Division ASCE* 116 (1990): 567–84.

CHAPTER 7

Water for the Environment: Updating Texas Water Law

Mary E. Kelly

*F*ish and wildlife need water. It's a simple truth, but ensuring that fish (including shellfish) and wildlife, and the habitats they depend on, have sufficient fresh water in the face of increasing—and shifting—human demand for water is a complex challenge. In Texas, as in many other western states that rely primarily on the prior appropriation doctrine (see Chapter 3), ensuring water for the environment presents difficult legal questions. These questions have been front and center in Texas water law for the last decade, paralleled by new initiatives in the scientific arena to better define freshwater flow needs for various instream and estuarine habitats.[1] (See Chapter 6 for an in-depth discussion of the importance of freshwater flows for Texas estuaries.) As noted by the 2004 report of a select group of scientists charged with examining environmental flow issues, "The question is not whether environmental flows are important and should be protected, but rather, how, when and where, and in what quantities should flows be reserved for environmental purposes in the state's rivers and streams and its bays and estuaries."[2]

This chapter begins with a brief overview of environmental flow provisions in Texas law prior to 2007. It then explores more recent developments on the policy and legal fronts, including a new environmental flow law enacted by the 80th regular session of the Texas legislature. These developments reflect a concerted effort on the part of many decisionmakers, water suppliers, conservation organizations, and others to ensure that as Texas grows, it will still have sufficient freshwater flows to sustain the healthy fish and wildlife populations that are so central to the state's natural heritage and economy.[3] The chapter concludes with a brief look at other water law topics that bear on environmental flow protection, including water reuse, amendments to existing surface-water permits, and groundwater management.

BACKGROUND

Prior to 1985, Texas law did not contain any specific statutory mandates for the water rights permitting agency to consider the possible effects of an application for new surface-water use, diversion, or storage on instream or estuarine habitat. The vast majority of existing surface-water rights were recognized, through proceedings to adjudicate legal rights,[4] or issued before 1985. Then, in 1985, the legislature amended Section 11.147 of the Texas Water Code to mandate consideration of the effects of surface-water permit applications on bays and estuaries and instream uses. Up until 2007, the relevant portions of that section provided as follows:

> Section 11.147. Effects of Permit on Bays and Estuaries and Instream Uses.
>
> (b) In its consideration of an application for a permit to store, take, or divert water, the commission [now Texas Commission on Environmental Quality] shall assess the effects, if any, of the issuance of the permit on the bays and estuaries of Texas. For permits issued within an area that is 200 river miles off the coast, to commence from the mouth of the river then inland, the commission shall include in the permit, to the extent practicable when considering all the public interests ... those conditions necessary to maintain beneficial inflows to any affected bay and estuary system. ...
>
> (d) In its consideration of an application to store, take or divert water, the commission shall include in the permit, to the extent practicable when considering all the public interests, those conditions considered by the commission necessary to maintain existing instream uses and water quality of the stream or river to which the application applies.
>
> (e) The commission shall include in the permit, to the extent practicable when considering all the public interests, those conditions considered by the commission necessary to maintain fish and wildlife habitats...

With this amendment, the legislature provided that the Texas Parks and Wildlife Department (TPWD) was to be given notice of permit applications and the opportunity to offer evidence and testimony as a party to any hearing on the application.[5] The Water Code also was amended to provide that 5% of the yield of any reservoir that might be built after 1985 with state financial participation and located within 200 miles of the coast would be appropriated to the TPWD for "use to make releases to bays and estuaries and for instream uses."[6]

In general, the application of these provisions to individual permit applications was to help mitigate against adverse downstream environmental impacts of increased storage, diversion, or use of water.[7] While a step in the right direction, only about 15% of the currently permitted surface-water rights by volume actually have been subjected to the 1985 requirements. On the Brazos River, for example, only 56 of 1,087 water rights have environmental flow conditions. Most of these conditions are merely a prohibition on diversion of water when the riverflow is below a given threshold value. Other examples of the application of the provisions include target releases from post-1985 reservoirs closer to the coast, such as Lake Texana and Choke Canyon.[8]

More fundamentally, the 1985 provisions did not require setting of environmental flow targets or examination of the cumulative effect of proposed storage, use, and diversions on a particular river system or basin, and thus their effect has been limited to a permit-by-permit analysis, which has not captured reality on the ground. In addition, the permit conditions under the 1985 provisions have generally been based on protecting a "minimum" streamflow, which often does not necessarily result in sufficient protection of fish and wildlife habitat.[9] More sophisticated understanding of river systems has shown that "flow regimes that vary over time create a mosaic of diverse habitat conditions that in turn support diverse floral and faunal assemblages. ... A varied flow regime that mimics natural historical patterns and that maintains adequate sediment loadings is the key to sustaining the fish and wildlife resources within and adjacent to Texas rivers and streams."[10]

Prior to 2007, Texas law did have a few other provisions reflecting legislative concern about protecting water for the environment. In 1997, as part of the groundbreaking omnibus water legislation known as Senate Bill 1, the legislature established the Texas Water Trust. The operation of the trust is discussed in more detail below, but the important point here is that it was created to hold water rights dedicated to environmental needs, including instream flows and freshwater inflows to bays and estuaries.[11] The legislature also required state agencies to undertake a variety of studies of bay and estuary and instream flow needs.[12]

Texas water law also contemplated the potential use of a "reservation" to protect environmental flows. Section 11.046 of the Water Code provides that a surface-water permittee can reuse "surplus water" before discharging it back to the stream, but once it is discharged, it becomes "subject to reservation for instream uses or beneficial inflows or to appropriation by others." To date, however, neither the Texas Commission on Environmental Quality (TCEQ) nor its predecessor agencies have used this provision to establish a formal instream flow reservation.

The importance of water for the environment also was recognized by the legislature in establishing the regional water-planning process in 1997. The legislature directed that the regional plans make appropriate provision for environmental water needs and for the effect of upstream development on the bays, estuaries, and arms of the Gulf of Mexico.[13] Views differ on the extent to which this directive has been implemented, but a full analysis of that debate is beyond the scope of this chapter.

Finally, a word about the connections, or lack thereof, between groundwater management and protection of environmental flows: in a number of Texas watersheds, flow from springs, which depend on groundwater, constitute a significant part of the base flow of streams and rivers.[14] At the state level, no review has been made of how groundwater withdrawals affect springflow and consequently instream flows. The implications of this gap are well illustrated by the situation in the Edwards Aquifer, where groundwater extraction has significant implications for flows at Comal and San Marcos Springs, home to endangered species, and for flows in the Guadalupe River system (for an in-depth examination of the Edwards Aquifer issues, see Chapter 5). As discussed in the last section of this chapter, local groundwater districts are only just beginning to consider the effect of groundwater withdrawals on springflows.

POLICY AND LEGAL DEVELOPMENTS

Over the last several years, environmental flow protection has become one of the central water law and policy issues in Texas. This situation was triggered in large part by the filing in 2000 of a new water rights permit application, seeking to appropriate essentially all of the remaining unappropriated flow of the Guadalupe River for environmental uses.[15]

The San Marcos River Foundation Application

In July 2000, the nonprofit San Marcos River Foundation (SMRF) filed a new water right application with the TCEQ for more than 1 million acre-feet of water in the Guadalupe basin. According to SMRF, the goal of the application was to secure instream flows and freshwater flows needed to maintain the ecological integrity of Guadalupe and San Antonio Bays.[16] This included the flows needed to maintain healthy populations of blue crabs, the primary food source of the endangered whooping crane. SMRF proposed that if the TCEQ granted the application, the foundation would transfer the water right to the TPWD for placement in the Texas Water Trust.

The SMRF application was based on TCEQ water right rules, adopted in 1999, which defined "beneficial use" as including "instream uses, water quality, aquatic and wildlife habitat, or freshwater inflows to bays and estuaries."[17] The commission further defined "instream use" as "the beneficial use of instream flows for such purposes including, but not limited to, navigation, recreation, hydropower, fisheries, game preserves, stock raising, park purposes, aesthetics, water quality protection, aquatic and riparian wildlife habitat, freshwater inflows for bays and estuaries, and any other instream use recognized by law."[18]

The TCEQ rules are more specific than Texas law, which does not expressly define "instream use" as a beneficial use. Instead, the statute defines "beneficial use" as including "any other beneficial use," in addition to specifically enumerated uses including for domestic, municipal, agricultural, mining, hydroelectric power, navigation, recreation and pleasure, public park, and game preserve purposes.[19]

Although TCEQ and its predecessor agencies had previously issued water rights permits that included instream use as part of a series of authorized uses, the SMRF application was essentially breaking new ground by requesting a large new appropriation solely for environmental and instream use. The TCEQ staff reviewed the application and, in late 2002, issued a notice of draft permit and setting the matter for a full hearing. The draft permit proposed 87,000 acre-feet for instream flows in the San Marcos River and 980,000 acre-feet for freshwater inflows to the Guadalupe and San Antonio Bay systems. Meanwhile, three other new applications for instream environmental water use had been filed. The Galveston Bay Preservation and Conservation Association requested appropriation for the Galveston Bay system; the Matagorda Bay Foundation requested a permit for freshwater inflows to the Matagorda Bay system; and the Caddo Lake Institute filed an application for environmental, recreational, and navigational purposes in the Caddo Lake system in East Texas.[20] Together, these appropriations would total millions of acre-feet for environmental purposes.

These applications generated significant interest and controversy.[21] Conservation, fishing, and recreation interests tended to view the applications positively, as a necessary step to protect environmental flows in lieu of a comprehensive and effective state approach and in the face of competing new applications for consumptive water uses. Water suppliers and state decisionmakers tended to take an unfavorable view of the applications. For example, in a 2003 amicus curiae brief to TCEQ, Lieutenant Governor David Dewhurst opined that the commission should defer action on the application until the conclusion of the legislative session. He asked that the agency give the legislature an opportunity to clarify the law because, in his opinion, it was not clear that the commission was authorized to issue a permit to appropriate state water to be used solely to maintain instream flows.

In March 2003, the three-member governing board of TCEQ dismissed the SMRF application, declining to send it through the full hearing process. The dismissal order stated that the "Commission did not have the express statutory authority to, nor should authority be implied from the Texas Water Code § 11.023, to issue the permit." The applications for Galveston and Matagorda Bays and Caddo Lake also were dismissed by TCEQ, in December 2003. The dismissal of these applications came after the legislature had decided to take up and pass legislation on the environmental flow issues. As discussed in more detail below, these dismissals were appealed to the courts.

Legislative Moratorium

The environmental flow applications triggered action by the 2003 session of the legislature, initiated by the filing of Senate Bill (SB) 1374.[22] While recognizing the importance of protecting instream and freshwater flows to maintain a "sound ecological environment," the legislation was designed both to place a temporary (two-year) moratorium on the issuance of new permits *solely* for instream flow for environmental purposes and to set up a study commission to provide recommendations on how to address environmental flow issues.

Specifically, the bill provided that the legislature had "not expressly authorized granting water rights exclusively for: (1) instream flows dedicated to environmental needs or inflows to the state's bay and estuary systems; or (2) other similar beneficial uses."[23] It went on to provide that no new permits for instream flow could be issued, but subjected that section to a two-year sunset provision.[24] Despite these provisions, the bill also expressly reaffirmed that existing surface-water rights could be converted to instream flow, and that instream flow could be added as an authorized use to an existing right. It also strengthened language in

Section 11.147 related to evaluating new permit applications for their effect on the environment.[25]

The bill passed the senate and the house Natural Resources Committee, but it failed to get a vote on the house floor because of legislative deadlines for house consideration of senate bills. In the last days of the regular session, however, the provisions of SB 1374 were tacked on to SB 1639, a bill dealing with groundwater issues, as it came to the floor of the house of representatives. Approved on a 131 to 8 vote, the amended SB 1639 was quickly approved by the senate and signed by the governor.

Environmental Flows Study Commission

The Environmental Flows Study Commission established by SB 1639 initiated its work in February 2004. The commission was composed of three state senators; three state representatives; the chairs of the TCEQ, TWDB, and TPWD; two academics; three river authority general managers; and one private practice attorney. In order to assist its work, the commission appointed a scientific advisory committee (SAC) at its first meeting.

The study commission held one hearing to collect information on how other states were dealing with environmental flow issues and take testimony regarding how Texas should approach the issue.[26] In a parallel process, the SAC also held hearings to gather scientific and technical information on environmental flow issues. The SAC issued its report in October 2004, providing a comprehensive overview of key scientific issues in protecting environmental flows in Texas, as well as several findings. In particular, the SAC found that because of the variation in climatologic, hydrologic, and aquatic environments across the state, a "one-size-fits-all" approach to protecting environmental flows was not appropriate for Texas. It also concluded that "participation by stakeholders and water interests in the environmental flow program and rigorous scientific review are of paramount importance to achieving acceptable environmental flows, that any process should incorporate adaptive management and precautionary principle methods," and that both regulatory and market mechanisms for protecting flows should be explored to provide a "more comprehensive and effective environmental flow program."[27]

However, as 2004 slipped away and the 2005 legislative session loomed, the study commission still had not issued recommendations. In an effort to break the gridlock and move forward, a small group of stakeholders was convened. The group included two members of the commission, representatives of the Texas Water Conservation Association (the state's major water supplier trade association), and representatives of the National Wildlife Federation, the Lone Star Chapter of the Sierra Club, and Environmental Defense.[28] After several weeks of discussion, the stakeholder group agreed on a consensus position, which was forwarded to the full study commission for consideration. In late 2004, the study commission adopted the consensus agreement as the basis of its report to the legislature, along with the SAC report.[29] The primary features of the consensus agreement, which were then turned into legislation for the 2005 session, reflected the central SAC findings of a basin-by-basin approach, good science, plenty of involvement of stakeholders, adaptive management, and a combination of regulatory and market approaches.

The 2005 Session

The recommendations of the Environmental Flows Study Commission were developed into detailed legislative language and included as Article 1 of SB 3, sponsored by Senator Ken Armbrister, then chair of the senate Natural Resources Committee.[30] The committee began hearings on SB 3 on April 12, 2005, one week after it was officially filed, but more than three months into Texas's five-month, biennial legislative session. Other provisions of this multitopic bill—such as those dealing with water infrastructure financing, groundwater, and the Edwards Aquifer—proved to be quite controversial, delaying the introduction of the bill, the committee's deliberations, and the committee vote. The legislation eventually reached the senate floor in late April. It was passed and sent to the house, where it was approved by the house Natural Resources Committee. The house committee made various changes to several components of SB 3, but the environmental flow provisions of Article 1 remained largely intact.

The bill was placed on the house calendar on May 24, the last day for house action on senate bills. However, the crowded state of the calendar and other priorities prevented a full house vote on SB 3. Attempts to add the environmental flow provisions of Article 1 to other bills in the last days of the regular session were unsuccessful.

Governor's Environmental Flows Advisory Commission

With the environmental flow issue still unresolved, in October 2005, Governor Rick Perry issued an executive order establishing a new Environmental Flows Advisory Commission (EFAC).[31] Members of the EFAC were appointed in March 2006. The EFAC was chaired by Rod Pittman, who was also chair of the TWDB, and included the chairs of the TCEQ and TPWD, as well as several public members appointed by the governor to represent a diversity of interests. The governor requested that the EFAC "examine relevant issues and make recommendations for commission action and legislation on methods for making future decisions to protect instream flows and freshwater inflows, while integrating such needs with human needs, including methods to address allocation of flows during drought conditions, using the December 2004 report of the Study Commission as a starting point."[32]

The EFAC held several hearings during 2006, examining whether any changes should be made to the provisions of Article 1 of SB 3 as it was filed in 2005. It also explored the use of markets to protect flows in other western states—that is, the voluntary conversion of existing agricultural or other rights to instream flow—as well the operations of the Texas Water Trust and land stewardship as a method for enhancing streamflows. The EFAC also appointed a science advisory group, whose report provided recommendations on what constitutes "sound ecological environment" and on strengthening the science of existing environmental flow study programs at state agencies. With respect to "sound ecological environment," the committee recommended the following definition:

A sound ecological environment is one that:

- sustains the full complement of native species in perpetuity;
- sustains key habitat features required by these species;
- retains key features of the natural flow regime required by these species to complete their life cycles; and
- sustains key ecosystem processes and services, such as elemental cycling and the productivity of important plant and animal populations.[33]

The EFAC's final report was issued on December 20, 2006. It contained 32 recommendations and adopted those from the science group. Some of the EFAC recommendations called for modest adjustments in the SB 3, Article 1 process, relating to such items as the composition of an oversight Environmental Flows Committee, the composition of bay/basin stakeholder committees, and collaboration between bay/basin science and stakeholder groups. The EFAC also suggested increased use of voluntary market mechanisms and incentives to foster the use of existing rights to meet environmental flow needs.[34]

The 2007 Session

In 2007, the environmental flow provisions were incorporated as Article 1 of the reworked SB 3 and filed separately in the house of representatives as House Bill (HB) 3. Both bills were sponsored by the respective Natural Resource Committee chairs, Senator Kip Averitt and Representative Robert Puente. SB 3 was again a multitopic bill, including provisions on reservoir site designation, water conservation, and the Edwards Aquifer. And once again, the fate of the bills was in doubt until the very last day of the session, due largely to controversy over provisions unrelated to environmental flows. This time, however, the result was different. Both SB 3, with the environmental flow provisions intact, and HB 3 passed both houses and were signed into law by the governor.

The key features of Article 1 of SB 3 and HB 3 include the following:[35]

- A new process for ensuring quality science and full participation of local and regional stakeholders in setting environmental flow targets and desired flow regimes. The legislation established bay/basin science committees, as well as diverse bay/basin stakeholder committees for each major river basin in Texas. The science group was directed to come up with environmental flow regime recommendations to protect a "sound ecological environment," and the stakeholder group was to review and comment on these recommendations, in consideration of other factors, including present and future water needs in the basin, and to explore ways the environmental flow targets can be met. The work of both committees was then to be forwarded to both the Environmental Flows Commission (for comment) and the TCEQ.

- Encouragement of creative solutions for integrating environmental flow protection and water supply development decisions to find approaches that meet both purposes. The legislation provided that TCEQ, via a rulemaking process, consider the bay/basin-specific science and recommendations and set appropriate environmental flow standards for each bay/basin system, adequate to support a "sound ecological environment." The law also directed TCEQ to establish a "set aside" of unappropriated water, if available, to help meet the flow standards. All permit requests for new appropriations of surface water would then have to comply with the flow standards. In consideration of the fact that several major applications for new surface-water appropriations were currently pending at TCEQ and could reach a decision point before the flow standards are enacted, the law also provided that the environmental flow conditions on permits for new appropriations issued between the effective date of the legislation (September 1, 2007) and the date the flow standards are finalized by TCEQ could be adjusted if necessary to help meet the flow standards, but it capped that adjustment at 12.5% of the flow in order to provide increased certainty to water rights permit holders.
- Legislative oversight of the process through the continuation of the Environmental Flows Commission. The law established an Environmental Flows Advisory Group (EFAG), consisting of three state senators, three state representatives, and one representative each from the TCEQ, TWDB, and TPWD. The EFAG would name bay/basin stakeholder committees, provide comments on their recommendations, and oversee the process.
- Clear timetables and process requirements for action to protect environmental flows and to provide greater certainty for water planning and permitting. The legislation contained specific deadlines for appointing bay/basin committees, for the committees to do their work, and for TCEQ rulemaking. In order to make the process manageable, the legislation provided for a staged approach. The first two basin/bay systems to be addressed would be the Trinity–San Jacinto–Galveston Bay system and the Sabine–Neches–Sabine Lake Bay system. The target for final TCEQ rule adoption for this first system is October 1, 2010, about three years after the effective date of the legislation. The second set includes the Colorado–Lavaca Rivers and associated bays and the Guadalupe–San Antonio–Aransas River and Bay system. The remainder of the systems would be addressed in the third tier. All told, the process would spread over about 10 years.
- Provisions for adaptive management by requiring periodic review and, if necessary, adjustment of flow standards, using the same process described above.
- Assurances that environmental flow requirements may be suspended during times of drought emergencies if necessary to provide water for other essential beneficial uses.
- Reaffirmation of the ability to voluntarily convert existing rights to instream use or add instream use to existing rights.

In sum, this groundbreaking new law reflected the principles embodied in the consensus agreement between water suppliers and three of state's major conservation groups. These principles include broad stakeholder participation in developing environmental flow regime targets;[36] a basin-by-basin approach that provided flexibility in setting environmental flow targets; the use of state-determined flow reservations instead of private applications for new environmental flow appropriations; an ability to adjust flow conditions on water use permits that might be issued before the flow regime recommendations could be finalized; and adaptive management over time to meet changing conditions and respond to enhanced scientific understanding of the relationships between flows and a "sound ecological environment" in streams, rivers, and bays.

Other Environmental Flows Activities

During the last few years the environmental flows legislation was being formulated and considered, some local efforts began to address flow issues. For example, the Galveston Bay Freshwater Inflows Group (GBFIG), which includes representatives of natural resource agencies, environmental groups, fisheries, agricultural interests, and water suppliers, has been studying Galveston Bay freshwater inflow needs since 1996 and strategies to meet those needs. It has been providing information and recommendations to the Region H Regional Water Planning Group on how to incorporate Galveston Bay freshwater flow needs into the regional planning process.[37] It is likely that GBFIG members will be active participants in the new environmental flows process, especially as Galveston Bay is part of the Trinity–San Jacinto basin, one of the first to be addressed by the new process.

Academics, water managers, resource agencies, conservation organizations, and others have also gathered to examine environmental flow needs for Caddo Lake in east Texas.[38] The group has met several times to explore data needs, begin to define target flows in the key tributaries to the lake, and identify ways those needs can be met while meeting demand for other uses.

Also, under a 2001 legislative directive, the TWDB, TPWD, and TCEQ have been conducting field studies and developing methodologies to establish instream flow needs for several priority basins.[39] The agencies have benefited from a National Academy of Sciences review of their methodology, as well as input from a variety of stakeholders.[40] However, the process has been underfunded and is currently behind the original schedule, which contemplated completion of the high-priority studies by 2010. Studies have been completed on the Lower Brazos and Sulphur Rivers and are in process on the San Antonio, Sabine, and Brazos Rivers.

Court Challenges

Parallel to the legislative process, SMRF and the other environmental flow permit applicants continued to pursue court challenges to TCEQ's dismissal of their environmental flow permit applications.[41] In February 2006, the Travis County district court granted SMRF's partial motion for summary judgment and remanded the case to TCEQ for further processing. The state appealed the order, and the Corpus Christi court of appeals overturned the district court, declaring the matter moot because of the passage of SB 3.[42]

It is beyond the scope of this chapter to discuss all the legal arguments surrounding these cases, but a few basic aspects are worthy of note. Essentially, the proponents argued that TCEQ regulations allowed such new applications for instream flows; nothing in the statute at the time the application was filed barred TCEQ from issuing such permits; and the legislature cannot retroactively ban the issuance of such a permit. In response, the attorney general of Texas, representing TCEQ, argued at the district court level that diversion or impoundment of surface water is an implied prerequisite for an appropriation of water for beneficial use under Texas law, and an application for instream use does not meet this prerequisite; and that the provisions of SB 1639 indicate that the legislature never intended for TCEQ to have authority to issue new water rights permits solely for instream, environmental flow purposes. These issues are similar to those that have arisen in other western states' battles over environmental flow protection. The July 2008 court of appeals decision did not directly decide these issues. It found instead that the provisions of SB 3 that prohibited the TCEQ from issuing a new water right permit for instream flows dedicated to environmental needs or bay and estuary inflows rendered the SMRF application moot.[43] SMRF appealed the decision to the Texas Supreme Court, but its petition for review was denied without comment in May 2009.

Role of Existing Water Rights

Apart from the issue of large new appropriations or set-asides of unappropriated water to protect environmental flows, many observers believe it also will be necessary to use market and other voluntary transactions involving existing surface-water rights to meet environmental flow targets.[44] The Texas Water Trust, created in 1997 in an effort to facilitate the use of existing rights to help meet environmental flow needs, and administered by the TWDB, can accept temporary or permanent donations of existing surface-water rights for use in protecting instream flows, water quality, fish and wildlife habitat, and freshwater flows to bays and estuaries.[45] The water right is protected from cancellation while it is in the trust. The dedication of rights to the trust must be reviewed and approved by the TCEQ (generally thought to require a permit amendment to authorize instream use), in consultation with the TPWD and TWDB. The Texas Department of Agriculture also may provide input to the review and approval process for dedication of rights to the trust.

To date, however, the trust holds only two water rights: one for 1,236 acre-feet in the Upper Rio Grande, donated in 2003, and one for 33,108 acre-feet in the San Marcos River, donated by Texas State University in 2006.[46] Unlike water trusts in other parts of the country, such as the Oregon and Washington Water Trusts, the Texas Water Trust does not have funding to purchase, lease, or otherwise acquire water rights, nor is it heavily promoted by TWDB. These limitations have been a significant constraint on the trust's effectiveness.

The 2006 EFAC recommendations emphasized the need to promote the use of voluntary market mechanisms to help protect instream flows. This approach was reflected in the provisions of SB 3/HB 3, including allowing deposits to the trust to be credited against any environmental flow standards that may ultimately be adopted by the TCEQ and reaffirming the ability to add instream use to existing water rights.

Three private water trusts also have been established recently in Texas:[47] the Guadalupe Blanco River Trust, established by the Guadalupe Blanco River Authority; the Trans-Pecos Water Trust, established by local ranchers and farmers, with assistance from conservation organizations, to protect flows in the Rio Grande and its tributaries; and the San Saba Trust, established by local landowners to protect flows in the San Saba. Each of these private trusts is in the early stage of operation. It remains to be seen whether they can lease, purchase, or secure donations of significant amounts of water rights (most likely unused or underused agricultural water rights) to protect or augment instream flows.

RELATED ISSUES

Developments in other areas of Texas water law also may affect the options and need for protecting environmental flows. Three areas are worthy of particular attention: water reuse, amendments to surface-water permits, and groundwater management.

Water Reuse

In some Texas river basins, much of the flow during drier times is made up of return flows from municipal, industrial, or agricultural operations.[48] This means that the use of these return flows has implications both for downstream water rights and for instream flows and bay and estuary inflows. As pressure on supply grows, many water providers are trying to get the most out of their existing rights by using the water completely and as many times as possible. For example, in a city that diverts water for municipal needs, not all the water will be consumed. Some will end up at the sewage treatment plant, usually on the order of 40% to 60%. The usual practice has been to discharge the treated sewage back to the river. Increasingly, however, cities are looking to reuse—or even sell—the treated effluent for green space irrigation, use in cooling towers, or other nondrinking consumptive uses.

Texas law allows for both indirect and direct reuse of water. Indirect reuse refers to discharging return flows into a watercourse and then removing them downstream for use; reuse without first discharging to a stream is known as direct reuse. Texas law does not require a permit for direct reuse, but a "bed and banks" permit is required for indirect reuse.[49]

According to the 2007 state water plan, reuse is projected to provide 13% of "new" supply in 2010 and up to almost 15% in 2050.[50] As of this writing, several large indirect reuse applications are pending at TCEQ, including proposals by the cities of Austin and Houston, San Jacinto River Authority, North Texas Municipal Water District, and Upper Trinity Regional Water District.[51]

Significant questions remain under current Texas law about when and how indirect reuse applications should be treated, and it is likely that these issues will be addressed in future sessions of the legislature. The 2004 SAC report suggested that it might be desirable to "reserve" 10% to 30% of return flows for environmental flow purposes, while noting that return flows originating from groundwater or interbasin transfers of surface water might require "special consideration," as they were not necessarily originally part of the basin's flow.[52] Because each potential reuse situation can be quite unique, it is difficult to find solutions that have broad applicability. However, the process established by SB 3/HB 3 would allow for stakeholders to consider whether and how to use return flows to help meet environmental flow targets on a basin- or bay-specific basis.

Amendments to Surface-water Permits

As water demand increases and demand patterns shift among various traditional permitted uses of surface water, water right transfers are gaining interest as a way to meet demand without construction of new reservoirs (see Chapter 4). In some cases, however, these transfers have implications for environmental flows. For example, if a permittee has been using only a portion of its authorized right, the instream habitat may have come to depend on the flows being passed through. If the permittee then proposes to transfer that unused portion of the right to another user in a different location, this could result in adverse effects on the stream or on downstream water users. In many western states, water right holders may transfer only the amount of water historically consumed under the permit.[53] In Texas, however, provisions of SB 1, enacted in 1997 as Section 11.122(b) of the Texas Water Code,[54] purported to allow transfer of the full "paper right" as long as the effects on downstream water right holders or the environment were no greater than if the authorized right had been fully exercised. Many had interpreted this provision as obviating the opportunity for downstream water right holders or others to request a contested case hearing on the transfer application. Others contended that a hearing was still required, even if the full right could be transferred, in order to determine whether conditions on the transfer were required to protect downstream interests or instream habitat.

These competing interpretations of the statute came to a head in the case of *City of Marshall v. City of Uncertain et al.*[55] The city of Marshall sought to transfer a portion of its 16,000 acre-feet of water right to a power plant for industrial use. Up to that point, Marshall had never used more than half of its full water right. Several groups sought a hearing on the proposed transfer, but the TCEQ denied the hearing request. The groups appealed the decision. The Travis County District Court reversed the TCEQ's decision, upon which the TCEQ appealed to the court of appeals, which upheld the district court.

The case then went to the Texas Supreme Court, which ruled "that section 11.122(b) does not mandate issuance of Marshall's water-rights amendment without the assessment of other substantive criteria imposed by the Water Code and the Commission's rules. ... [P]ersons affected by these substantive criteria are entitled to notice and hearing to determine the proposed amendment's effect, or ... the Commission could determine ... a hearing is not necessary. We believe that the Commission should make this determination in light of our construction of section 11.122(b)." This is another area where legislative action could occur in the future.

Groundwater Management

State water law in Texas provides little in the way of direct links between groundwater management and base flows in streams and rivers, although such links exist in many basins. However, as the "desired future condition" of various aquifers begins to be defined through the groundwater management area (GMA) process required under HB 1763,[56] these linkages could be strengthened if, in appropriate circumstances, the desired future condition is set to include maintenance of springflows and seep flows that contribute to base flows in streams and rivers.[57] This process is still in the early stages, and it remains to be seen how it will be implemented by the various GMAs; how the information will be used in regional water planning; and whether the desired future conditions set by the various GMAs will face challenges.[58] An outstanding question in some areas of the state is whether sufficient science is available on which to base a thorough review of the links among groundwater withdrawals, springflows, and surface-water flows.

CONCLUSIONS

With the 2007 enactment of SB 3/HB 3, the state of Texas now has one of the country's most ambitious and comprehensive environmental flow protection programs. Most policymakers and stakeholders have recognized that the biological and geographic diversity of the state, along with widely varying patterns of water availability, use, and demand, require that a successful environmental flow approach be flexible and basin-specific, as well as provide room for reasonable adaptive management. The challenge now is to fulfill the promise of SB 3/HB 3 by efficient and effective implementation. If that happens, the prospects are good that Texas can avoid the expensive gridlock and conflict that other states have faced when river management and water use clash with protection of endangered species, and instead leave healthy and thriving rivers and bays to future generations.

Even with the new environmental flow law, however, as water use patterns change and better science leads to new policy tools, there will be a continuing need to modernize other aspects of our state water law, especially those relating to reuse, permit amendments, and groundwater management.

NOTES

1. Science Advisory Committee (SAC), Report on Water for Environmental Flows, 2004, http://www.twdb.state.tx.us/EnvironmentalFlows/pdfs/SAC%20FINAL%20REPORT_

102704.pdf (accessed October 26, 2004); National Academy of Sciences (NAS)/National Research Council (NRC), *The Science of Environmental Flows: A Review of the Texas Instream Flow Program* (Washington, D.C.: National Academies Press, 2006); E.G. (Rod) Pittman, chair, Texas Water Development Board, Austin, TX, letter to Honorable Kenneth Armbrister and Honorable Robert Puente, co-presiding officers, Study Commission on Environmental Flows, Austin, TX, December 15, 2004.

2. SAC, Water for Environmental Flows.

3. Freshwater inflows to bays and estuaries are critical to the health of coastal commercial and recreational fisheries, which annually generate about $2 billion in economic activity (Larry McKinney, Why Bays Matter: Texas Needs Bays and Texas Bays Need Fresh Water, *Texas Parks and Wildlife* (July 2003): 24). Healthy rivers, wetlands, and estuaries also contribute to the state's growing ecotourism sector.

4. For a discussion of surface-water adjudication in Texas, see Texas Commission on Environmental Quality (TCEQ), *Rights to Surface Water in Texas*, Publication No. GI-228 (Austin, TX: TCEQ, 2009).

5. Texas Water Code, § 11.147(f).

6. Texas Water Code, § 16.1331. Subsection (e) of that section makes it clear that the provision is not intended to be in lieu of any other method of provided flows for instream uses.

7. SAC, Water for Environmental Flows.

8. Ibid.

9. Brian R. Richter, R. Matthews, D.L. Harrison, and R. Wiginton, Ecologically Sustainable Water Management: Managing River Flows for Ecological Integrity, *Ecological Applications* 13 (2003): 206–24.

10. SAC, Water for Environmental Flows.

11. The Texas Water Trust provisions are codified in § 15.7031 of the Texas Water Code.

12. See Texas Water Code, § 16.058 (bay and estuary studies) and § 16.059 (instream flow studies). Detailed information on these studies is beyond the scope of this chapter, but it can be found on various agency websites, including http://www.twdb.state.tx.us (Texas Water Development Board) and http://www.tpwd.state.tx.us (Texas Parks and Wildlife Department). Accessed September 13, 2010.

13. Texas Water Code, § 16.053(e)(5)(F).

14. SAC, Water for Environmental Flows.

15. Interestingly, that same year, a conservation task force established by former governor George W. Bush found that "management of water is probably the single most critical conservation issue in Texas." Governor's Task Force on Conservation, Taking Care of Texas, 2000, 35, available at www.tpwd.state.tx.us/publications/nonpwdpubs/media/taking_care_of_texas_report.pdf (accessed December 15, 2005).

16. See the San Marcos River Foundation website, http://www.sanmarcosriver.org (accessed December 15, 2006).

17. Texas Administrative Code, § 297.43(10).

18. Ibid., § 297.1(25).

19. Texas Water Code, § 11.023. A beneficial use is further defined by statute as "use of the amount of water which is economically necessary for a purpose authorized by (Chapter 11 of the Water Code), when reasonable intelligence and reasonable diligence are used in applying water to that purpose." Texas Water Code, § 11.002(4). Surface water can be permitted only for beneficial use.

20. Navigation and "recreation and pleasure" are both statutorily authorized beneficial uses.

21. Jake Bernstein, Does a River Have a Right to Flow? *Texas Observer*, June 21, 2002.

22. SB 1374 was introduced in the 78th Session of the Texas legislature (regular session), available at www.capitol.state.tx.us (accessed December 15, 2006).

23. Added as § 11.0235(d) of the Texas Water Code.

24. The moratorium, codified in § 11.01237 of the Texas Water Code, expired on September 1, 2005.

25. SB 1639, § 3, amended § 11.147 (d) and (e) of the Texas Water Code, as follows (previous language appears in strike-through):

> (d) In its consideration of an application to store, take, or divert water, the commission shall include in the permit, to the extent practicable when considering all public interests, those conditions considered by the commission necessary to maintain [consider the effect, if any, of the issuance of the permit on] existing instream uses and water quality of the stream or river to which the application applies.
>
> (e) The commission shall include in the permit, to the extent practicable when considering all public interests, those conditions considered by the commission necessary to maintain [also consider the effect, if any, of the issuance of the permit on] fish and wildlife habitats.

26. Joint Committee on the Study Commission on Water for Environmental Flows, Interim Report to the 79th Texas Legislature, December 2004.

27. SAC, Water for Environmental Flows.

28. The author represented Environmental Defense in these discussions.

29. Joint Committee, Interim Report.

30. SB 3 also contained articles regarding land stewardship, groundwater management, water conservation, infrastructure financing and management of the Edwards Aquifer, among other topics. It followed in the tradition of two previous omnibus water bills, SB 1 in 1997 and SB 2 in 2001.

31. Office of the Governor of Texas, Governor Perry Creates Environmental Flow Advisory Commission, October 28, 2005. For detailed information on the hearings and recommendations of the EFAC, see http://www.twdb.state.tx.us/EnvironmentalFlows/index.html (accessed December 15, 2006).

32. Ibid.

33. SAC, Recommendations of the Science Advisory Committee to the Governor's Environmental Flows Advisory Committee, 2006, http://www.twdb.state.tx.us/Environmental Flows/pdfs/Meeting5/SAC_Recommendations.pdf (accessed August 21, 2006).

34. Texas Water Development Board (TWDB), Environmental Flows Advisory Committee (EFAC) Final Report, December 2006, http://www.twdb.state.tx.us/Environmental-Flows/pdfs/REPORT/EFAC_FINAL_REPORT.pdf (accessed April 17, 2008).

35. For a full section-by-section analysis of Article 1 of SB 3, see the Senate Research Center Bill Analysis, 2007, http://www.capitol.state.tx.us/tlodocs/80R/analysis/html/SB00003F.HTM (accessed December 15, 2007). For a more general description of the flows legislation and the implementation process, see Texas Water Matters, The Environmental Flows Allocation Process, http://www.texaswatermatters.org/flows.htm (accessed December 15, 2007).

36. This aspect of the process is based loosely on the stakeholder-driven, adaptive management approach to environmental flow protection developed by The Nature Conservancy. Richter et al., Ecologically Sustainable Water Management.

37. For more information on GBFIG, see http://galvbaydata.org/WaterandSediment/FreshwaterInflowsGroupGBFIG/tabid/217/Default.aspx (accessed June 8, 2010).

38. For more information, see the Caddo Lake Institute website, http://www.caddolakeinstitute.us (accessed December 15, 2006).

39. For detailed information about the instream flow study process, see the Texas Instream Flow Program website, http://www.twdb.state.tx.us/InstreamFlows/index.html (accessed December 15, 2006).

40. NAS/NRC, *Science of Environmental Flows*. See SAC, Water for Environmental Flows, for additional discussion of the state's instream flow program.

41. Myron J. Hess, *Environmental Flow Protection in Texas* (presented at the 15th Annual Texas Environmental Superconference, Austin, TX, August 7–8, 2003).

42. *San Marcos River Foundation v. Texas Commission on Environmental Quality*, 267 S.W.3d 356 (Tex. App.–Corpus Christi, July 31, 2008).

43. Ibid.

44. Steven Malloch, Liquid Assets: Protecting and Restoring the West's Rivers and Wetlands through Environmental Water Transactions (Trout Unlimited, 2005), available at http://cbwtp.org/jsp/cbwtp/library/documents/Liquid%20Assets.pdf (accessed November 2008).

45. Texas Water Code, § 15.7031.

46. Colette Barron, In Water We Trust, *Texas Parks and Wildlife* (July 2006).

47. Ibid.

48. SAC, Water for Environmental Flows.

49. § 11.046 of the Texas Water Code allows reuse of water before discharge to the stream (direct reuse), but once it is in the stream, the water cannot be reused without a "bed and banks" permit under § 11.042 (indirect reuse). § 11.042 (c) provides for consideration of the requested reuse on "instream flows and freshwater inflows to bays and estuaries," but it offers little guidance on how that issue should be considered or what type of permit conditions might be authorized to avoid such adverse effects.

50. TWDB, *Water for Texas 2007* (Austin, TX: TWDB, 2007).

51. See details at TCEQ, Permitting, http://www.tceq.state.tx.us/permitting/water_supply/water_rights/pending.html (accessed June 8, 2010).

52. SAC, Water for Environmental Flows.

53. Terry L. Anderson and Pamela Snyder, *Priming the Invisible Pump* (Bozeman, MT: Property and Environment Research Center, 1997).

54. § 11.122(b) provides: "Subject to meeting all other applicable requirements of this chapter for the approval of an application, an amendment, except an amendment to a water right that increases the amount of water authorized to be diverted or the authorized rate of diversion, shall be authorized if the requested change will not cause adverse impact on other water right holders or the environment on the stream of greater magnitude than under circumstances in which the permit, certified filing, or certificate of adjudication that is sought to be amended was fully exercised according to its terms and conditions as they existed before the requested amendment."

55. Tex. Sup. Ct.–June 9, 2006.

56. Enacted in 2005 and codified in various sections of ch. 36, Texas Water Code.

57. See 31 Texas Administrative Code, § 356.2(8) for the regulatory definition of "desired future condition." See Robert Mace, R. Petrossian, R. Bradley, and W.F. Mullican, A Streetcar Named Desired Future Conditions: The New Groundwater Availability for Texas (presented at the State Bar of Texas 7th Annual Changing Face of Water Rights in Texas, San Antonio, TX, May 18–19, 2006), for more discussion of the changes in groundwater management brought about by HB 1763.

58. Mace et al., Streetcar Named Desired Future Conditions.

CHAPTER 8

Texas Boundary Water Agreements

Kathy Alexander Martin

*T*he boundaries of natural resources, watersheds in particular, frequently do not coincide with political boundaries. Furthermore, the distribution of water resources among political subdivisions is rarely equal or unanimously agreed on. This can lead to conflict at both the interstate and international levels. A mechanism for resolving these types of conflicts at the interstate level in the United States is interstate compacts. At the international level, nations that share water resources often agree to allocation and joint administration of these resources through treaties between the national governments. The border waters of Texas are subject to both compacts and international treaties. Examination of the Texas experience provides evidence of both the difficulties of achieving agreement on water resource issues in the first place and subsequently administering those agreements in light of changing circumstances.

Texas shares significant water sources, both surface water and groundwater, with several other states: Colorado, New Mexico, Louisiana, and Oklahoma. The United States also shares certain waters with Mexico, both being sovereign nations. The U.S. share of these waters is either used by federal projects or allocated by Texas in accordance with state statutes. This chapter provides an account of the interstate compacts between Texas and other states. It also discusses the two international water treaties—one in 1906 and the other in 1944—involving the United States and Mexico. These treaties have a profound impact on Texans residing in the border region.

A brief commentary on each compact traces the formative events and discusses the issues and legal disputes that continue to affect these relationships. The two compacts and two international treaties involving allocation of the Rio Grande are then examined in more detail, noting that the lessons learned from the long history of conflict and cooperation in the Rio Grande basin can have broad applicability to other international river basins. Relevant details include the historical evolution of these interstate and international agreements, the status of unresolved issues hampering their administration, and initiatives for managing the river's water in the future.

INTERSTATE COMPACTS

As with other western states, increasingly fierce disputes over shared water resources between Texas and its neighbors led to negotiations, culminating in agreements known as interstate compacts. In addition to efforts by the individual states, the federal government frequently encouraged settlement of such disputes by conditioning project subsidies on the ability of the involved states to finalize water allocation agreements. One deficiency of tying compact ratification to federal funding, however, was that the most contentious issues frequently were not addressed, leading to future litigation. In addition, Texas's interstate water compacts and treaties cover only surface-water allocations; groundwater was not explicitly included. The effect of increased groundwater pumpage on surface-water flows, however, was very important in disputes over water resources in the Pecos River basin. One likely reason for failure to include groundwater is the different legal framework for its administration. (See Chapter 3 for an in-depth discussion of the differences between groundwater and surface-water management regimes.) Texas's interstate compacts either addressed water quality issues at the outset or incorporated them at a later date. Water flows supporting aquatic species and other ecosystem uses were not originally included in Texas's interstate compacts, although these issues are currently being addressed through litigation under the Endangered Species Act or updated agreements between signatory states as part of compact administration.

Compact administration typically rests with a commission consisting of one or more representatives from each state and a nonvoting federal representative. In Texas, compacts are incorporated into state law, with the governor appointing commissioners for six-year terms. Decisions of the compact commissions must be made by unanimous vote, making decisions over controversial issues, particularly water allocation issues, difficult to resolve. When unanimity cannot be reached, disputes may be referred to the U.S. Supreme Court.[1] The high cost of litigation may tend to keep compact participants at the table. Requirements for unanimous decisions also allow recalcitrant states to thwart resolution of contentious issues in favor of the status quo.

Texas participates in the following compacts with the listed states (see Figures 8.1 and 8.2):

- Canadian River Compact: New Mexico and Oklahoma.
- Red River Compact: Oklahoma, Arkansas, Louisiana.
- Sabine River Compact: Louisiana.
- Pecos River Compact: New Mexico.
- Rio Grande Compact: New Mexico.

The United States also signed a water treaty with Mexico regarding the Rio Grande, which determined the amount of water available for Texas water users.

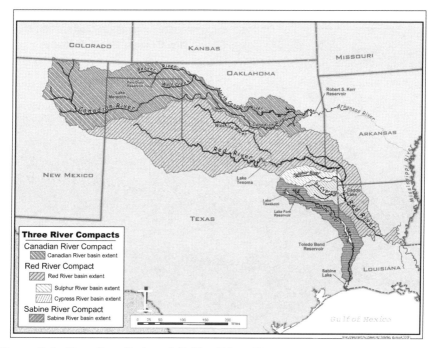

Figure 8.1 *Canadian, Red, and Sabine River Compacts*

Source: Map by Prescott Christian.

CANADIAN RIVER COMPACT

The Canadian River (see Figure 8.1) is a tributary of the Arkansas River, rising from its headwaters in the Sangre de Cristo Mountains in southern Colorado, near the state's boundary with New Mexico, then flowing south and southeast through New Mexico and Texas before entering Oklahoma and continuing to its mouth at the Arkansas River. The need to allocate the waters of the Canadian River was recognized in the early twentieth century. The Canadian River Development Association was formed in 1925 to begin work on flooding issues and irrigation projects, leading to an effort to negotiate a compact in 1926. Although the legislatures of New Mexico and Oklahoma ratified the compact, the Texas legislature failed to do so, with the result that this compact did not take effect.[2]

In 1938, the U.S. Army Corps of Engineers (USACE) received authorization from Congress to construct Conchas Reservoir in New Mexico for flood control purposes, thereby benefiting the basin states, and to provide irrigation water in New Mexico.[3] By the 1940s, declining water levels in the Ogallala Aquifer caused Texas farmers and cities to push for construction of Lake Meredith, in order to use the heretofore "wasted" surface waters of the Canadian River.[4] The combination of Texas's existing and future uses of surface water in the region, New Mexico's desire to protect existing rights and retain their ability to construct

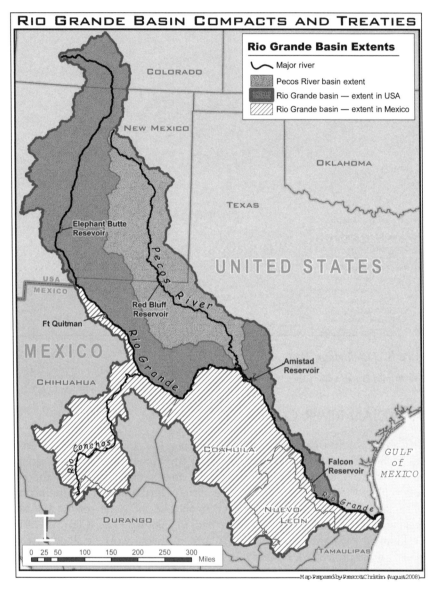

Figure 8.2 *Rio Grande basin compacts and treaties, including Pecos River Compact*

Source: Map by Prescott Christian.

additional storage in the future, and Oklahoma's desire to lock up water for future projects[5] led to congressional approval for the three states to begin negotiation of the Canadian River Compact.[6] Texas, Oklahoma, and New Mexico signed the compact on December 6, 1950, and the U.S. Congress approved it in 1952.[7]

Provisions of the Compact

The Canadian River Compact is somewhat different from Texas's other compacts with respect to the manner in which the river water is apportioned. This compact contains no requirements for delivery of specified amounts of water to the state line. Instead, it allocates water to each state by limiting the amount of water New Mexico and Texas can hold in conservation storage in their reservoirs.[8] Free and unrestricted use of the water is allowed in Oklahoma, as the downstream state. The requirements for Texas and New Mexico are more complex.

New Mexico is entitled to all waters originating above Conchas Dam, as well as all waters originating below the dam subject to storage limitations.[9] The compact specifies circumstances under which New Mexico and Texas may impound water in excess of their described conservation storage limits.[10] Texas is entitled to free and unrestricted use of the water within the state, subject to storage limitations. Its use of the north Canadian River is subject to usage constraints.[11] Texas is limited to the impoundment of 500,000 acre-feet of water, subject to certain conditions relating to Oklahoma's impoundment of water.[12]

Post-Compact Administration

After the compact took effect, several disputes arose among the signatory states over interpretation of its requirements. Two disputes, relating to interpretation of Article V(b), were resolved via a resolution of the commission and memorialized by an agreement between Texas and Oklahoma in 1981.[13] The third dispute required judicial resolution in the U.S. Supreme Court.

Once the Canadian River Compact was signed, Congress authorized the Canadian River Project to construct Lake Meredith on the Canadian River in Texas, which was completed in 1964. In the meantime, New Mexico constructed Ute Reservoir, located below Conchas Dam, in 1963, and subsequently enlarged this reservoir in the early 1980s. Texas and Oklahoma objected to this enlargement because of its effects on their own downstream interests and took the dispute to the Supreme Court, arguing that the terms of the compact limited New Mexico's storage below Conchas Dam.[14]

Events in 1987, while the case was pending, exacerbated the dispute. A flood in the river above Conchas Dam resulted in a spill of approximately 250,000 acre-feet of water. New Mexico impounded 60% of this water in Ute Reservoir and passed only 40% downstream to Texas and Oklahoma. New Mexico asserted that compact limitations did not apply to the spilled waters because those waters originated above Conchas Dam. Texas and Oklahoma disagreed and amended their original complaint.[15] The court appointed a special master, who investigated the complaints and filed a report on October 15, 1990.[16] A judgment and decree entered on December 13, 1993, held that New Mexico had been in violation of the compact and ordered water releases from Ute Reservoir and that the state pay attorney's fees.[17]

Although there have been no recent court filings, issues over compact interpretation continue. Oklahoma contends that Texas is in violation of Article

V(a) of the compact. Texas constructed Palo Duro Reservoir, which is used for recreational purposes and stops the flow of Palo Duro Creek as a result of its design.[18] Texas contends that Palo Duro Reservoir was constructed for municipal purposes and complies with the terms of the compact. The states discussed this issue in the 1990s, and in 2001, the Oklahoma legislature adopted a resolution calling for that state's attorney general to proceed with a lawsuit over the issue, although no suit has yet been filed as of this writing.[19]

RED RIVER COMPACT

The Red River (see Figure 8.1) is part of the Mississippi drainage, rising from headwaters in New Mexico, flowing through the Texas Panhandle, and becoming the border between Oklahoma and Texas. The river then flows through Arkansas and into Louisiana, where it empties into the Atchafalaya and Old Rivers. The basin drainage in Texas also includes the Sulphur River and Cypress Creek basins.[20]

Negotiations over the equitable apportionment of the waters of the Red River basin in Oklahoma, Texas, Arkansas, and Louisiana began in 1956, in response to the prolonged drought of the 1950s. Issues included water development (Texas and Oklahoma), shortages of water for industrial use (Arkansas), and flood control (Louisiana). The compact took more than 20 years to negotiate because of disagreements between Texas and Oklahoma, which were ultimately resolved in 1976. The U.S. Congress consented to the Red River Compact on December 22, 1980.[21]

Provisions of the Compact

The Red River Compact includes provisions unique among Texas's interstate compacts. The U.S. Supreme Court typically has original jurisdiction in suits between states. This compact allows U.S. district courts concurrent original jurisdiction in any disputes over compact interpretation.[22] In addition, it contains provisions expressly related to water pollution.[23] The compact does not require annual accounting for enforcement purposes at this time, but the states could agree to this in the future.[24] The states may incorporate water accounting as uses increase and water supply development continues.[25]

Because of the size of the basin and the involvement of multiple states, the apportionment of water flows is somewhat complex. The basin is divided into five reaches, three of which relate to Texas. In Reach I, flow in the interstate tributaries is apportioned 60% to Texas and 40% to Oklahoma. Intrastate streams are apportioned solely to the state in which they are located. The waters of the mainstem are divided equally between Texas and Oklahoma, including flow in the river and storage in Lake Texoma.[26]

In Reach II, Texas has free and unrestricted access to intrastate streams in Texas. For the mainstem of the Red River below Denison Dam and tributaries except for those already described in other subbasins, the four states have equal rights to the waters as long as specific flow requirements are met at the Arkansas-Louisiana border. No state is required to guarantee a minimum flow to any other compact state.[27]

In Reach III, which includes the Texas portion of streams crossing the Texas-Arkansas boundary and flowing into Cypress Creek – Twelve Mile Bayou watershed in Louisiana, the waters are apportioned 60% to Texas and 40% to Arkansas. For tributaries crossing the Texas-Louisiana boundary and flowing into Caddo Lake, Cypress Creek – Twelve Mile Bayou, or Cross Lake, Texas and Louisiana each have free and unrestricted access to flows originating within the particular state, subject to requirements related to water inflows into and use of Caddo Lake.[28]

Post-Compact Administration

No litigation has occurred to date among the states regarding interpretation of compact provisions. However, issues exist that potentially could affect relationships among the signatory states, particularly Texas and Oklahoma. Disputes between Texas and Oklahoma over the waters of Sweetwater Creek and the North Fork of the Red River predate the compact. In April 2008, this dispute was resolved with the adoption of administrative rules. The new resolution states that Texas did not violate the compact, and that water flows in both streams would be divided 60% to Texas and 40% to Oklahoma.[29]

A current controversy involves a Dallas-area entity's desire to divert water in Oklahoma. In January 2007, the Tarrant Regional Water District (TRWD) requested a permit from the Oklahoma Water Resources Board to divert 460,000 acre-feet of water from three basins in Oklahoma. However, once the water enters the mainstem of the Red River, it becomes saline. Thus, TRWD wants to divert the water in Oklahoma before it enters the river. In 2002, Oklahoma imposed a moratorium on water sales until completion of a 50-year water plan.[30] TRWD filed suit against the Oklahoma Water Resources Board and Oklahoma Water Conservation Storage Commission, challenging the moratorium and alleging its right to apply for, and be granted, a permit pursuant to the Red River Compact and the Commerce and Supremacy Clauses of the U.S. Constitution. The district court ruled that the lawsuit could continue, with the 10th Circuit Court of Appeals upholding the decision in 2008.[31] To further complicate the issue of Texas's diversions of Oklahoma water, the city of Irving, Texas, entered into a contract with the city of Hugo, Oklahoma, and filed suit challenging the moratorium. Irving's actions fueled disputes among the Texas entities seeking Oklahoma water.[32] Three of the major entities signed a historic agreement to share any water from Oklahoma, including the costs of infrastructure associated with the transfer.[33]

SABINE RIVER COMPACT

The Sabine River (see Figure 8.1) originates in northeast Texas and flows southeasterly through Panola County, where the river forms the boundary between Texas and Louisiana. The Sabine then flows into Sabine Lake, at the confluence of the Sabine and Texas's Neches River, and then through Sabine Pass into the Gulf of Mexico. As with most interstate compacts, the Sabine River Compact arose in response to competing claims for water. Louisiana claimed title in 1949 to all of the water below the point where the river becomes the boundary between Texas and Louisiana. Subsequent negotiations between the two states proceeded rapidly, with congressional approval in 1951, the onset of negotiations in 1952, and congressional ratification of the compact in 1954.[34]

Provisions of the Compact

The Sabine River Compact recognizes existing uses in both states, with any water withdrawals being subject to water availability, as determined by the compact.[35] The stateline reach, that portion of the river beginning at the point where it becomes the boundary between the states and ending in Sabine Lake, is apportioned equally between the two states.[36] The compact also requires both states to use water flows as they occur,[37] and it allows both states to use the bed and banks of the Sabine River to convey stored water without losing ownership of that water.[38] Water uses in both states are subject to the maintenance of a minimum flow of 36 cubic feet per second (cfs) at the state line.[39]

Post-Compact Administration

To date, all disputes between the two states have been resolved through negotiations. In fact, Texas and Louisiana cooperated in the construction of Toledo Bend Reservoir, said to be the only public water conservation and hydroelectric project in the United States built without federal funding.[40] Current issues include ongoing efforts in Texas to determine ecological instream flow uses in the Sabine River, which has caused concern among water users in Louisiana. Local entities there are monitoring the Texas instream flow process.[41]

OVERVIEW OF RIO GRANDE BASIN AGREEMENTS

The Rio Grande and Pecos Compacts, as well as two international agreements between the United States and Mexico, govern allocation of water flow in the Rio Grande (see Figure 8.2). The river drains two countries, the United States and Mexico, and eight states within these countries: Texas, New Mexico, Colorado, Chihuahua, Coahuila, Durango, Nuevo Leon, and Tamaulipas. The Rio Grande is the fifth-longest river in North America, with a drainage area of approximately 355,000 square miles. The basin covers 11% of the continental United States and 44% of the land area in Mexico.[42]

The Rio Grande headwaters originate in the San Juan Mountains of Colorado. The river enters the San Luis Valley of southern Colorado, where it has supported extensive irrigated agriculture since 1851.[43] After entering New Mexico, most of the river's flow is diverted for agricultural purposes, with a large portion of this use being managed through various federal projects administered by the U.S. Bureau of Reclamation (BOR). After passing through central New Mexico, the river is impounded by Elephant Butte Reservoir. Elephant Butte stores water for irrigation uses in New Mexico and farther downstream in Texas, as well as providing delivery of 60,000 acre-feet of water in Mexico for irrigation use, pursuant to the Convention of 1906.

At a point approximately 23 miles north of El Paso, the river enters Texas and becomes the boundary between the United States and Mexico. The river at this point is severely degraded and frequently dries up during the winter months. Intensive agricultural diversions by irrigation districts on both sides of the border have resulted in severe alteration of the natural water flow regime of the river. For example, the average annual water flow at Fort Quitman was reduced 96% between 1884 and 1964.[44] The river then flows through a reach known as the "Forgotten River" from below Fort Quitman to Big Bend National Park, located at the confluence of the Rio Grande and Mexico's Rio Conchos. Because of upstream diversions and the construction of Elephant Butte Reservoir, the river's flow is often reduced to zero at times when its flow would have occurred historically. When the Rio Grande reaches Presidio and Ojinaga, flow from the Rio Conchos in Mexico replenishes the mainstem of the river. Historically, the Rio Conchos contributed approximately two-thirds of the flow of the Rio Grande below its confluence. This water contribution declined over the last 10 years, however, leading to water disputes between the United States and Mexico.

The river then enters a reach extending from Amistad Reservoir to the Gulf of Mexico. The Pecos and Devil's Rivers, both tributaries on the Texas side, enter the mainstem of the Rio Grande at Amistad. The river is heavily regulated in this segment, with water quality being an issue. Pressures on the river system are primarily a function of explosive population growth on both sides of the border. The population of the Lower Rio Grande Valley on the Texas side, for example, is at least 1.2 million people, and the population on the Mexican side of the border has been estimated to be at least three times greater.[45] The total population could exceed 11 million people in 2030. In addition to high population growth rates, the border region is one of the poorest in the United States, with the border counties on the Texas side having the lowest per capita income in the country. Most river water in the Lower Rio Grande is used for irrigation. The amount of irrigated land, however, is steadily declining as a result of urbanization.[46]

Water supply historically has been an issue in the Rio Grande basin, generating conflict at the local, state, and international levels. Although land use in the basin is primarily agricultural, rapidly urbanizing areas are stressing the water resource base. Basin problems include habitat loss and endangered species management, water quality degradation, and water management issues resulting from shifts in water usage from agricultural to municipal. Although numerous programs address basin problems, fragmented authority at all governmental levels and a reluctance to address water quality concerns constitute ongoing binational challenges.

The international character of the Lower Rio Grande basin introduces added complexity, as national foreign policy interests are involved. Thus, control over the

U.S. share of the waters in the river vests at the federal level, with those interests and allocation of the waters of the main channel being managed by the U.S. section of the International Boundary and Water Commission (USIBWC). The USIBWC assists in determining the flow in the river and apportioning the national shares between the United States and Mexico. Once the national flows are allocated, the remaining water flow is available for Texas users in accordance with state statutes. Hence the actual water flow allocated in any given year and issues that could arise if treaty provisions are violated are outside the control of the state. Because of this reality, the interests of Texas water users are subordinate to the national interests of the United States.

THE 1906 CONVENTION

Water supply problems in the Rio Grande basin were apparent by 1878. Texan complaints of excessive Mexican diversion in 1888, plans by the United States to construct a storage reservoir in the El Paso area in 1890, and complaints by Mexico over upstream diversions in the United States in 1894 all led to a joint investigation of water supply by the two countries, beginning in 1896. The International Boundary Commission (IBC) investigated the feasibility of constructing an international reservoir above El Paso. Mexico supported construction of the reservoir, claiming monetary damages resulting from increased U.S. diversions upstream and suggesting that the United States should fund construction of the reservoir to settle this debt. In November 1904, at the 12th International Irrigation Congress in El Paso, representatives of Texas, New Mexico, and Mexico reached a compromise, endorsing the Elephant Butte dam site.[47]

Once the parties reached a tentative agreement, several obstacles still remained. The BOR would construct the project, but the Reclamation Act of 1902 did not authorize construction of projects in Texas.[48] Numerous federal agencies and other interests supported construction of the reservoir, and Congress approved construction in 1905 and extended the Reclamation Act to cover Texas in June 1906.[49] U.S. concerns about allocating water to Mexico, however, threatened to delay construction of the project. In addition, during negotiations, Mexico requested a survey of irrigable lands, different water allocation amounts, and division of water flows below El Paso. Despite Mexico's concerns with the proposed treaty, it withdrew its requests, U.S. interests agreed to a defined allocation to Mexico, and the 1906 Convention was ratified by the two countries.[50]

The purpose of the 1906 Convention was equitable distribution of the waters of the Rio Grande for irrigation purposes, as well as removal of all causes of controversy with respect to its distribution.[51] The 1906 Convention provided for the delivery of 60,000 acre-feet of water per year to Mexico on completion of the Elephant Butte Dam.[52] Article II provided for delivery of this water in the same proportions as water delivered to U.S. users in El Paso. In the case of extraordinary drought, however, deliveries to Mexico would be reduced in the same proportion as deliveries to U.S. users. This method of allocation provides for risk sharing between the two countries during droughts. The remaining articles state the following:

- that the United States incurs all costs of storing, delivering, and measuring deliveries of water in the bed of the river at the head of the Mexican canal;
- that Mexico waives all claims of damages for previous delivery shortages; and
- that Mexico relinquishes any claim to waters of the Rio Grande between El Paso–Juarez and Fort Quitman, Texas.[53]

PECOS RIVER COMPACT

The Pecos River (see Figure 8.2) is a major tributary of the Rio Grande, originating in the Santa Fe Mountains in New Mexico, flowing south and entering Texas near the 104th meridian, and then flowing into the backwaters of the Rio Grande at Lake Amistad.[54] The river suffers from large water withdrawals for irrigation, as well as increased groundwater pumping in New Mexico's Roswell area.[55] Additionally, the Pecos is naturally saline, and salt cedar infestations throughout its watershed have increased salinity even further by transpiring much of the available water. Large-scale irrigation projects on the Pecos River in Texas date back to the mid-1800s on the tributaries and the late 1800s for mainstem projects.[56] Because of these dams, irrigation projects, and groundwater pumping in New Mexico, water flows in the Pecos River are substantially reduced, with estimates of these reductions being as high as 99% of the flow.[57]

Efforts to negotiate a compact between Texas and New Mexico began in 1923 to facilitate Texas's plans to construct Red Bluff Reservoir, located downstream of the state line. The compact was signed by the states but subsequently vetoed by the governor of New Mexico.[58] New Mexico's interest in developing Alamogordo Reservoir led to a compromise in 1935, but the state again did not ratify the agreement, continuing to develop groundwater in the Roswell area. The issue of groundwater development in New Mexico, as well as problems with salinity, led to the Pecos River Joint Investigation under the auspices of the federal government. On conclusion of the investigation, and armed with data relating to water use and supply, salinity, and flooding, Texas and New Mexico once again began compact negotiations in 1942 and finally reached an agreement in 1948.[59]

Provisions of the Compact

The Pecos River Compact allocates waters of the Pecos River between Texas and New Mexico, based on a formula summarized as follows:

- New Mexico shall not deplete the flow of the Pecos River at the state line below a quantity equivalent to that available to Texas under 1947 conditions.
- Texas receives the flow of the Delaware River.
- Water salvaged in New Mexico is allocated 43% to Texas and 57% to New Mexico.

- Unappropriated floodwaters are allocated 50% to Texas and 50% to New Mexico.[60]

Salvage water is additional water in the stream resulting from eradication of salt cedar in New Mexico. The "1947 conditions" noted in the compact are based on the results of investigations conducted in support of the negotiations, which led to the development of an *Inflow-Outflow Manual* outlining the quantity of water New Mexico should deliver under varying circumstances.[61]

Post-Compact Administration

Within a year after the compact was signed, issues arose over how to account for each state's water use and interpretations of compact provisions relating to the 1947 condition.[62] Deliveries to Texas pursuant to the compact were substantially less than should have occurred based on the *Inflow-Outflow Manual*.[63] Texas filed suit against New Mexico, requesting that deliveries be made in accordance with the terms and conditions of the compact, and the court appointed a river master to determine New Mexico's compact delivery obligations with this system still in place.[64] In addition, the court required New Mexico to pay Texas $14 million for previous compact underdeliveries of water.[65] Since the litigation, New Mexico has tried numerous times to revise the methods used to determine the quantity of water it is required to deliver to Texas. To date, Texas has not been in agreement with these revisions. New Mexico also has taken steps to retire existing water rights and augment river flows with groundwater to meet its delivery obligations.[66]

In addition to disputes over water deliveries, compliance with the Endangered Species Act (ESA) complicates administration of the compact. One year after settlement of the lawsuit over water allocations, the U.S. Fish and Wildlife Service (FWS) listed the Pecos bluntnose shiner as a threatened species under the ESA. Water flow regimes necessary to maintain this fish require a continuous flow at lower rates, leading to increased evaporative loss in a river where water quantity has been the predominant concern. Solutions to the problem include additional purchases of water rights and, according to some, renegotiation of the compact under the theory that water flows necessary to maintain the shiner serve a national interest. Under this theory, Texas would share the burden of the additional allocation with New Mexico.[67] Texas's position with respect to these issues is that compact deliveries must be protected.[68]

RIO GRANDE COMPACT

The jurisdiction of the Rio Grande Compact begins at the headwaters of the river in Colorado, extends through New Mexico, and ends at Fort Quitman, Texas (see Figure 8.2). Early water disputes in this region involved irrigators in New Mexico and below El Paso–Juarez in the United States and Mexico. The 1906 Convention and construction of Elephant Butte Dam partially resolved the dispute. The convention, however, did not address water allocations farther upstream in the

Middle Rio Grande Valley, and even farther upstream in the San Luis Valley in Colorado. Resolution of the dispute between upper and lower water appropriators began in 1896, with the U.S. Department of the Interior enacting an embargo on the use of public lands for water diversions from the Rio Grande or its tributaries in Colorado and New Mexico.[69] The purpose of the embargo was to protect the water yield of Elephant Butte Reservoir and ensure that water obligations to Mexico under the 1906 Convention were fulfilled.

Success in interstate compact negotiations over Colorado River water encouraged New Mexico and Colorado to begin discussions over Rio Grande water allocations. At the time, these states were opposed to the participation of Texas in the talks, mainly because it might give water users below Elephant Butte Reservoir more power in determining the negotiation outcome. Many upper basin users also felt that the inclusion of Texas would lead to calls for allocating water as far downstream as Brownsville, thereby also opening the door for participation by Mexico.[70] Texas eventually did join the negotiations, however.

By 1922, there were calls for lifting the Rio Grande embargo, with a study by the BOR concluding that additional upstream water storage would not substantially reduce the yield of Elephant Butte Reservoir. In 1925, the Interior Department approved a reservoir in Colorado. Texas and New Mexico were outraged, with many calling for Supreme Court adjudication of rights to the river's water. In addition, the secretary of the interior lifted the embargo, stating that it was an improper exercise of federal authority, in that water management within a state should be the responsibility of that state. New Mexico withdrew from compact negotiations and began preparing a lawsuit. The time and money required to mount such a suit, however, encouraged the state to return to the negotiation table. Negotiations began, and a temporary compact was signed in 1929 (the 1929 Compact), allowing talks to continue until 1935.[71]

Negotiations progressed slowly, and as a result of drought conditions in the 1930s, Texas filed suit against New Mexico in 1935 to force the water allocation issue.[72] One effect of this suit was federal reinstatement of the Rio Grande embargo because of jurisdictional conflicts among the federal agencies charged with administration of the river.[73] The states agreed to extend the 1929 Compact until 1937 to allow for a federal study of the Upper Rio Grande. The results of this joint investigation provided the data needed to determine water allocations among the states. The Rio Grande Compact Commission assigned the task of ironing out compact details to the engineer advisors, with the states subsequently reaching agreement and signing the Rio Grande Compact of 1938.[74]

Provisions of the Compact

The Rio Grande Compact allocates water flows from the headwaters in Colorado through New Mexico to Elephant Butte Reservoir to the three states, based on a formula that can be summarized as follows:

• Deliveries of water from Colorado to New Mexico are measured at the state line and are based on water flows in the Rio Grande and one of its tributaries,

according to a defined schedule. Any additional water pumped into the Rio Grande for purposes of meeting delivery requirements can be credited to Colorado only if the water meets specific water quality parameters.
- New Mexico's deliveries of Rio Grande water to Texas are measured at Elephant Butte and based on water flows at index gages.
- Determinations of water delivery shortages and credits are based on a formula specified in the compact, which contains provisions to determine methods of repayment of water shortages. The compact also contains provisions limiting water storage in reservoirs constructed after 1929 in Colorado and New Mexico, based on credit amounts.[75]

Post-Compact Administration

Compact operations began in 1940. New Mexico accrued water shortages of 331,800 acre-feet by 1951, and Texas filed suit. The Supreme Court appointed a special master, but the court dismissed the case in 1957 because the United States was not a party.[76] Meanwhile, the region was in the midst of an extreme drought, with the water debt of both Colorado and New Mexico mounting. New Mexico began efforts to increase the water flows delivered to Texas. Colorado continued to use more water than authorized, however, and New Mexico and Texas sued that state in 1966 for violation of the compact. The litigation was suspended when Colorado agreed to begin making water deliveries to meet its obligations. It also agreed to curtail water diversions within its boundaries to ensure that it met its compact obligations. Elephant Butte Reservoir spilled in 1985, subsequently relieving Colorado of its water debt. Colorado remains in compliance to the present time.[77]

As in the Pecos River, endangered species issues are assuming a growing role in compact administration. The FWS listed the Rio Grande silvery minnow as an endangered species in 1994.[78] The agency subsequently designated a reach of the Rio Grande in New Mexico as critical habitat for the fish in 1999. Maintenance of habitat for the silvery minnow has changed the manner in which New Mexico and the BOR manage the dams in the middle reaches of the Rio Grande in New Mexico. Keeping the water flow constant has eliminated several management options previously used to reduce water evaporation and other losses associated with water releases from upstream reservoirs for delivery downstream. Texas is working with New Mexico on this issue, having agreed to allow the other state to relinquish accrued credit water according to a schedule such that New Mexico could store water upstream of Elephant Butte to meet flow requirements for the silvery minnow.[79]

THE 1944 TREATY BETWEEN THE UNITED STATES AND MEXICO

Inadequate water supplies have been an issue in the Lower Rio Grande Valley for more than 100 years. Settlement in the valley began after the Civil War, primarily because of the attraction of the fertile soils in the Rio Grande delta. The first efforts to manage water supply were primarily allocative, focusing on quantification of the water available to both the United States and Mexico.

As settlement of the area intensified, periodic water scarcity and the highly variable flows of the then unregulated river made it clear that the full potential of this region would not be realized without storage for water supply and flood control.[80] Efforts to obtain federal assistance began in 1902, with a petition presented to the U.S. secretary of state requesting that an agreement be reached with Mexico on the distribution of the waters of the Rio Grande between its confluence with the Devil's River and the Gulf of Mexico. At that time, however, the State Department was involved with issues surrounding competing claims to waters of the Rio Grande upstream at El Paso and did not address water allocation issues in the lower basin.

In 1909, the Rio Grande Commission was established to study water allocation in the Lower Rio Grande. The two U.S. commissioners were also considering water apportionment issues and claims for allocation of Colorado River water, and thus consideration of the allocation of Rio Grande waters was not a priority for them. Further, the Mexican position on this issue was that the United States should not get one drop of water from the Rio Grande until all Mexican uses were satisfied. Texas farmers in the Lower Valley also were reluctant to begin negotiation of any treaty for allocation of Rio Grande water until Texas had developed as much irrigation use as possible, under the premise that any treaty would protect existing uses.[81]

Funds were appropriated to the International Boundary Commission (IBC) between 1910 and 1920 for studies of the Rio Grande, although no studies subsequently were generated. Irrigation development on Mexican tributary streams, however, contributed to increased attention to Rio Grande issues.[82] Congress established the Commission on Equitable Use of Water in the Rio Grande below Fort Quitman in December 1924.[83] Mexico refused to cooperate unless discussions on the Colorado River also were included. The United States agreed in 1927 to consider Colorado River issues, participating in the creation of a joint commission to address Rio Grande, Colorado River, and Tijuana River issues. The commissioners agreed to gather data on stream flows, water diversions, acreage under irrigation, possible future water uses, and flood control for the three rivers. The results of this study for the Lower Rio Grande established that the population in the region was 200,000, that 350,000 acres were under irrigation, that Mexican tributaries contributed 70% of the flow below Fort Quitman, and that Texas was using 70% of the flow of the river for irrigation.[84]

The Mexican and American representatives held competing positions on allocation of water flows in the Rio Grande. The parties could not reach agreement on the substantive issues at that time, and both sides agreed to further study. Irrigation by Texas farmers in the Lower Rio Grande Valley was indeed threatened at that time as it was increasing. On the Texas side of the border, 370,000 acres were under irrigation, while 4,300 acres were irrigated on the Mexican side. Further, Mexico had begun constructing dams on the

Rio Conchos, Rio Salado, and Rio San Juan and had spent large sums on irrigation works.[85]

The negotiation process moved slowly, accomplishing little until 1932, when the American section of the International Water Commission merged with the USIBC. The Mexican government was instituting agrarian reforms during this period, including redistribution of land and socialization of agriculture.[86] Mexico also constructed a gravity canal, the Retamal, capable of diverting the entire low flow of the Rio Grande. The United States initiated studies in 1936 of water use in the Rio Grande, to form a basis for planning water storage facilities. The study expanded in 1938 to include flood control and water conservation. By 1940, 583,000 acres were under irrigation in Texas, with water supply becoming critical to continued development in the region. The 1938 study advocated construction of an American canal and two off-channel water storage facilities to ensure a firm supply to U.S. irrigators in the Lower Valley.[87]

From the 1920s until the early 1940s, Colorado River issues overshadowed negotiations on the Rio Grande, effectively preventing any progress toward an equitable distribution of the latter's waters. As politicians wrangled over Colorado River issues, however, a crisis situation developed on the Rio Grande. Mexico proposed development of El Azucar Dam in 1937, with the goal of forcing resolution of water allocation issues for the Colorado River. Further, a flood occurred in the region, followed by a drought so severe there was inadequate water supply for domestic use. Water for municipal use, for example, was shipped into Brownsville and Matamoros.[88] Saline irrigation return flows contributed to water quality problems in the river, and some thought that construction of reservoirs would alleviate the problem by diluting the saline waters.[89]

The Texas Board of Water Engineers presented a resolution to the U.S. State Department in 1938, endorsed by 152 irrigation districts, service clubs, chambers of commerce, and municipal users, urging action on Rio Grande issues and construction of water storage facilities.[90] In response to construction of Mexico's Retamal project, the United States proposed construction of an American canal upstream of the Mexican diversion facility. Threats from both sides of the border finally forced resolution of water allocation issues involving the Colorado River and Rio Grande, and the United States and Mexico signed a treaty in 1944 dividing the waters of the two rivers between them. At the time of the treaty, no precedent existed in international law regarding the apportionment of a boundary river with irrigation use on both sides.[91]

Provisions of the Treaty

The area governed by treaty allocations encompasses the mainstem of the Rio Grande and its tributaries, from Fort Quitman to the Gulf of Mexico (see Figure 8.2).[92] The 1944 Treaty established an allocation priority for joint water use, beginning with domestic and municipal use, and ending with other beneficial uses as determined by the International Boundary and Water Commission (IBWC). The treaty did not explicitly mention environmental flows, although fishing and hunting were included, but with a low use priority.[93] The treaty assigned water flows between the United States and Mexico as follows:

• Mexico would receive all water from the San Juan and Alamo Rivers, including return flows; 50% of the flow below the lowest major storage reservoir (Falcon Reservoir), as long as that water was not already allocated; 66% of the flow from six measured Mexican tributaries, including the Rio Conchos, subject to certain conditions; and 50% of all other water flows, including ungaged tributary inflows occurring between Fort Quitman and the Amistad Reservoir.
• The United States would receive all waters of the Pecos and Devils Rivers, Goodenough Spring, and Alamito, Terlingua, San Felipe, and Pinto Creeks; 50% of the water flow below the Falcon Reservoir that was not already allocated; 33% of the flow of the six measured Mexican tributaries, provided such flow was not less than an annual minimum flow of 350,000 acre-feet in any five-year accounting period; and 50% of tributary inflows occurring between Fort Quitman and the Amistad Reservoir.[94]

The terms of the treaty included construction of shared storage reservoirs.[95] It also added oversight of accounting for water resources to the IBC, creating the International Boundary and Water Commission (IBWC). The IBWC includes a Mexican section, Comision Internacional de Limites y Aguas (CILA), and a U.S. section, USIBWC.[96] The IBWC is currently responsible for measuring Rio Grande waters, allocating these waters between the United States and Mexico, as well as flood control, water quality, and sanitation issues.

The treaty assigned the IBWC authority to handle disputes arising from interpretation and application of the treaty, subject to the approval of the governments of the signatory nations. In the event of a lack of agreement by the commissioners of the two sections, treaty provisions called for both commissioners to inform their respective governments, and the two governments then to pursue resolution of the dispute through diplomatic channels and other means pursuant to treaties between them.[97] The 1944 Treaty stipulated that the commissioners record any decisions in the form of minutes, which are essentially clarifications or interpretations of treaty provisions. Further, both governments would have what is essentially a veto power over any minutes agreed to by the commissioners. In the event that one of the governments disagreed, the two nations would negotiate an agreement, with the agreement then communicated back to the commission.[98] The treaty also provided that in the event of extraordinary drought that prevents Mexico from meeting delivery requirements of tributary waters, deficiencies could be made up in the next five-year accounting cycle.[99]

The IBWC coordinates with the other federal entities to participate in sanitation projects and assists with annual scheduling of deliveries of 60,000 acre-feet of Rio Grande–Rio Bravo water to Mexico, in accordance with the Convention of 1906.[100] In addition to allocating Rio Grande waters, the treaty included provisions for allocating the waters of the Colorado and Tijuana Rivers.[101] Allocations of Colorado River water are germane to any discussion of Rio Grande issues, because any evaluation of fairness of the Rio Grande water allocations under the 1944 Treaty must consider the parallel issue of Colorado River water allocations.

The fact that the treaty linked allocation of the waters of both rivers helps ensure that Mexico will remain at the negotiating table to address any issues. At the time the treaty was signed, the two nations were involved in ongoing disputes over water allocation for both river systems. Users of Colorado River water, particularly California, were reluctant to acknowledge any Mexican rights to the river's water out of fear that such recognition would substantially reduce their own water allocations. Users of the Rio Grande in Texas were concerned that substantial investments and historical irrigation uses in the Lower Rio Grande Valley would be irreparably harmed if the river waters were divided fifty-fifty.[102]

As previously mentioned, Mexican tributaries, particularly the Rio Conchos, provide most of the water flow to downstream users in the Lower Rio Grande Valley. The agreement on the division of border waters essentially stipulated a quantifiable amount of Colorado River water to be delivered to Mexico, while protecting water users in the Lower Rio Grande Valley by apportioning a quantifiable amount of tributary inflow from the Rio Conchos and other Mexican tributaries to the United States for use in Texas. The minimum quantity of an average of 350,000 acre-feet per year of tributary water over a five-year period prevented Mexico from fully developing tributary waters in such a way as to keep sufficient water from reaching downstream users in the United States.

Post-Treaty Administration: Texas's Share

After the 1944 Treaty was signed, conflicts among water users in Texas became violent, and the Texas Supreme Court intervened.[103] The court held that lands adjacent to a stream carried implied rights of irrigation. In spite of the court's ruling, issues relating to Rio Grande riparian water rights remained unresolved.[104]

As a result of Rio Grande water shortages during the drought of the 1950s, various lawsuits were filed, with the Texas Supreme Court taking control of the administration of the state's use of the waters of the Rio Grande.[105] By the 1950s, the population of the Lower Rio Grande Valley on the U.S. side had increased to 450,000, and more than 700,000 acres were under irrigation.[106] Falcon Dam, jointly constructed by the United States and Mexico, was completed in 1954, enhancing the ability of Texas users to obtain water supplies.[107] Shortly thereafter, an extreme rainfall event filled the reservoir to capacity. The U.S. share of Falcon Lake water storage at that time was about 1.3 million acre-feet.

By January 1956, less than two years after it filled, the U.S. share declined to less than 700,000 acre-feet. Water storage continued to decline until, by June, only 50,000 acre-feet remained.[108] Because of declines in the U.S. share of Lake Falcon water storage, the Texas Board of Water Engineers determined that water remaining in storage would be limited to releases for domestic and municipal uses. Various legal suits and countersuits were filed. Of special concern was the idea that if more water were released than could be used, the excess water would flow into the Gulf of Mexico, thereby being "wasted."[109] The lawsuits included a request for the court to determine the quantities of water to be allocated to the competing users, as the available water from the U.S share of the Rio Grande was sufficient to cover only 50% of water uses. A final determination of water rights was essential for the future economic well-being of the Lower Rio Grande Valley, with the court apportioning the waters of the Lower Rio Grande among the various users.[110]

A watermaster program was created in the Lower Rio Grande Valley, pursuant to the rulings in the valley lawsuits described above. This was not the first time water use in the valley was allocated by a watermaster. In addition to the special master that allocated the U.S. share among water diverters during the several years of court proceedings, the users also attempted to create an informal watermaster program themselves. Because of problems caused by diversion of all water flows by upstream users, many valley irrigation districts signed an agreement in 1953, known as the Falcon Compact. Although it created an allocation mechanism, it was doomed to failure because compliance was voluntary.[111] Today the Rio Grande watermaster, appointed by the executive director of the Texas Commission on Environmental Quality (TCEQ), is responsible for day-to-day operations, accounting, and compliance.[112] As water enters the system, the IBWC allocates the shares, based on a formula that includes apportioning to Mexico the quantity of water granted by the 1944 Treaty. The watermaster allocates Texas's use of the U.S. share, based on the quantity of water stored in the Falcon-Amistad reservoir system and inflows into the reservoirs. Texas's water is then divided among certain reserves. Each water right owner is allotted an "account" against the system storage.

At the time of the original lawsuit, the first 125,000 acre-feet of inflows were dedicated to the domestic, municipal, and industrial (DMI) reserve. This volume has since increased to the first 225,000 acre feet of inflows. The DMI reserve is determined by rule.[113] After allocation of water flow to the DMI reserve, any water remaining in the irrigation accounts is reserved for their future use. (See Chapter 3 for a description of water rights in the Rio Grande.) The watermaster then creates an operating reserve, based on remaining water storage and including deductions for evaporation, channel losses, and any other water losses. Any water remaining after the above allocations is placed in the irrigation accounts. If there are additional water inflows during the accounting period, this water also is allocated to the irrigation accounts. The watermaster rules provide for a modified version of "use it or lose it." If an irrigator does not use the allocated water in his or her account within a two-year period, the account is reduced to zero and is not restored until the water right holder notifies the watermaster that he or she intends to begin using the water.[114]

Although the watermaster rules do not explicitly mention water marketing, the rules pertaining to water contracts allow some flexibility in water administration in the Middle and Lower Rio Grande. In other parts of the state, if a water right owner wants to change the place of use, diversion point, or diversion rate of his or her water, a permit amendment is required.[115] In the Middle and Lower Rio Grande, the water right owner need only file a copy of the contract with the TCEQ.[116] The only requirements are that the contract must be for purposes of use already authorized by the water right, and if not, a permit amendment is needed, although such amendments are treated as administrative in nature in the Rio Grande and are processed rapidly by the TCEQ. The absence of the amendment process for the Lower Rio Grande is justified by the region's unique hydrology and water rights framework. (See Chapter 4 for a detailed discussion of water marketing in this region.)

Even after years of litigation and negotiation, resulting in the creation of a system for managing water rights in the Lower Rio Grande, water shortages still occur. The 2006 Regional Water Plan for the Rio Grande indicates serious challenges, with predicted water shortages in all areas by 2050. The population projections in the plan suggest that the population in the Lower Rio Grande will triple over the next 50 years. The plan also identifies reductions in firm yield of the Amistad-Falcon system over the planning horizon due to sedimentation, estimating a decrease in firm yield of about 10%. Because of pro-marketing rules used to allocate shares of the U.S. portion of reservoir storage, irrigation users will absorb water shortages caused by this decrease.[117]

Another major concern is reduced tributary inflows from Mexican tributaries. Uncertainty over volumes of water available from Mexico pursuant to the treaty and a repayment schedule for deficits directly affect water planning in the Lower Rio Grande. Because of the presumed resolution of allocation issues embodied in the 1944 Treaty and the accounting procedures resulting from the Lower Rio Grande Valley lawsuits, water supply calculations should be straightforward. This is due to the strict state and international accounting schemes incorporated into these instruments. However, treaty compliance issues and absence of a definition of "extraordinary drought" hamper the ability of planners to project the quantity of water available under various hydrologic conditions, particularly drought, which is an important consideration for future planning.

Post-Treaty Administration: International Issues

Partially as a result of a severe drought beginning in the late 1990s, a dispute arose between the two nations over Mexico's delivery of tributary waters, as required by the 1944 Treaty. Mexico's position was that Article IV of the treaty allowed diminished water deliveries in times of extraordinary drought. The IBWC and the two governments resolved disputes as they arose during the extended drought and continued to negotiate when changes in circumstances warranted. The process, however, led to acrimony among local stakeholders on both sides of the border. Local politicians as well as the Mexico media called for renegotiation of the treaty during this period.[118]

This dispute began as a consequence of the onset of a basinwide drought in the early 1990s. Not only did the drought result in reduced rainfall and streamflows (hydrologic drought), but also inefficient irrigation practices and actions of water managers on both sides of the border (human-caused drought) may have contributed to the severity and duration of drought impacts. This combination led to the failure of Mexico to meet delivery requirements of tributary waters during the 1992–1997 accounting cycle. Specifically, the United States claimed that Mexico failed to meet delivery requirements of an annual minimum flow of

350,000 acre-feet of water from the Rio Conchos and the other Mexican tributaries. Mexico claims its water deliveries were reduced in accordance with treaty provisions relating to extreme drought.

There is little debate over whether drought conditions existed. Both the United States and Mexico issued drought disaster declarations numerous times throughout the 1990s and early 2000s.[119] Aside from the lack of a precise definition of extraordinary drought in the 1944 Treaty, which in itself has contributed to the present controversy, the treaty also does not adequately specify a repayment scenario in the event that treaty obligations are not met. The treaty states that any deficiencies in deliveries from one five-year accounting cycle should be made up in the next five-year cycle.[120] It does not, however, address how deficits should be met in the event of a drought lasting longer than five years. Specifically, what is to be done when, in the five-year period following a drought, Mexico is unable to deliver the current five-year amount in addition to the quantity of water necessary to satisfy the previous shortage?

Mexico's position was that water deliveries from all the measured tributaries and reallocation of water in storage in Amistad and Falcon Reservoirs were sufficient to meet its deficits incurred during the 1992–1997 accounting cycle, and that any deficiencies incurred during the 1997–2002 cycle were repayable during the 2002–2007 cycle. The U.S. position was that the treaty obligates Mexico to satisfy underdeliveries from the 1992–1997 cycle, as well as any deliveries required during the 1997–2002 cycle, during the 1997–2002 cycle.[121]

When the 1997 accounting concluded, Mexico proposed a method to repay deficits in that accounting cycle by releasing tributary waters when the flows exceeded a specified level. The United States rejected the offer and requested information on conditions in the tributaries. The two sides then began a series of technical meetings.[122] After two years of negotiation and an interim agreement transferring ownership of some water storage in treaty reservoirs from Mexico to the United States, the two nations signed Minute 307 on March 16, 2001. This minute called for water deliveries from Mexico based on rainfall projections, with assignment of specified quantities of water from Mexican tributaries.

Minute 307 also mentioned water releases from Carranza Dam in Mexico, as well as opening the possibility of releases from other Mexican reservoirs.[123] Local interests in Mexico strenuously objected to water releases from Carranza Dam, asserting that such releases would affect the local economy. The national government ignored the local interests, however, and Mexico released Carranza water to partially fulfill its obligations pursuant to Minute 307.[124] In addition to specifying methods for Mexican repayment of water deficits, Minute 307, recognizing the "spirit of friendship" between the two countries and their joint desire to prevent a recurrence of the events leading up to the agreement, committed the two nations to "identify measures of cooperation on drought management and sustainable management" of the Rio Grande basin.[125]

As a result of Mexico's increasing inability to comply with water delivery terms under the treaty, and as part of the cooperation called for in Minute 307, representatives of the two nations, in conjunction with their commissions, issued a joint memorandum agreeing to tour Mexican dams followed by another meeting in Austin. They also agreed to form a binational work group, for the purpose of initiating a summit on sustainable management in the basin, and to exchange data and ensure public access to IBWC data.[126]

Minute 308, signed June 28, 2002, contained provisions for assignment of Mexican water storage in Amistad and Falcon Reservoirs to the United States.[127] This agreement also included comments from the Mexican government that it intended to finance modernization of irrigation operations in the Rio Conchos watershed, with a goal of passing the conserved water downstream to the United States in order to reduce Mexico's treaty deficits. Minute 308 reiterated the two governments' support for the sustainability initiatives in Minute 307.[128] Minute 308 mentioned the formation of an International Advisory Council, composed of governmental and nongovernmental organizations, to act as a forum for exchange of information and provision of advice to the two countries,[129] but the formation of such a council was not included in the minute's recommendations. Minute 308 did include recommendations on agreement between the two governments on collection and exchange of drought-planning information, as well as having a binational summit of experts to address sustainable management.[130] Another agreement, Minute 309, was signed on July 3, 2003, and detailed proposed savings from water conservation initiatives.

Although the diplomatic negotiation process embodied in the IBWC minutes signed during the drought crisis did represent progress toward long-term resolution of basin issues, Texas water users excluded from the process were unhappy with the outcome. The U.S. section of the IBWC, although a federal agency, previously had been very responsive to border states' congressional delegations, whereas the IBWC's foreign policy component more closely resembled the domestic policy of the affected states.[131] This situation may be changing, however. During the water dispute, Texas farmers appealed to members of their congressional delegation, the Texas governor, the State Department, and George W. Bush, then President of the United States, all to no avail.[132]

As a result of the lack of response from elected officials, 17 Texas irrigation districts along with 16 individuals and 13 other entities (which together are hereinafter referred to as the districts), filed a claim under Chapter 11 of the North American Free Trade Agreement (NAFTA), alleging that Mexico's failure to deliver treaty water was a "taking," and asked for financial compensation of $500 million.[133] During the NAFTA proceedings, the TCEQ submitted a letter stating that claims of individual water users were not within the scope of negotiations over the water debt between the United States and Mexico. Mexico's position was that the claim fell outside the scope of NAFTA. The U.S. submission in the case also stated that the claims were outside NAFTA's jurisdiction. Federal intervention likely resulted from fear that a favorable ruling in the water case could affect a ruling in a similar type of case against the United States.[134] Other Texas politicians filed a letter in support of the claim.[135]

One issue brought before the arbitration panel was whether or not the water was a "good in commerce." The claimants made the argument that the Rio Grande was no longer a free-flowing river, and that the river's water was bought, sold, and traded; thus the water was a good in commerce. Mexico's position was that the tributary waters were subject to Mexican law, were not the property of the United States and, therefore, were not in commerce.[136] The Arbitration Tribunal found that NAFTA Chapter 11 did not apply in this case, meaning that the tribunal had no jurisdiction. NAFTA rules allow the judicial appeal of an arbitration ruling, and the districts appealed the ruling to a court in Canada, which upheld the arbitration panel's ruling.[137]

As outlined in the preceding section, the current dispute resolution process consists of negotiation by the U.S. and Mexican sections of the IBWC in order to reach a decision. The decision is then incorporated into the treaty as a minute. Any substantial changes to the treaty would require approval by the legislative bodies of the two countries. Previous discussion of the negotiations over allocations suggests that the minute system works fairly well and is sufficient to resolve conflicts as they occur. During times when municipal water use in Mexico was threatened, for example, the United States expressed concern for human water uses of the Mexican people, as evidenced in Minutes 240, 293, and 308. Minute 240, issued in response to drought affecting the municipal water supply of Tijuana, temporarily altered established water allocation mechanisms embodied in the treaty. Minute 293 contained provisions ensuring that municipal water uses were fulfilled in the Rio Grande basin. Minute 308 recognized Mexico's minimum water uses for human populations.[138]

Although the outcomes of the dispute resolution processes incorporated in the 1944 Treaty do indicate a history of resolving conflict and maintaining relationships between the two countries at an international level, relationships between state and local stakeholders suffer as a result of continuing disputes over interpretation of treaty provisions. This is due partly to shortcomings in the current dispute resolution process. One individual commenting on western water disputes recommended that basin processes be evaluated on the basis of the efficiency and fairness of the process and the durability of the outcome.[139]

In regard to efficiency (a process element), the first measure of evaluating negotiated agreements, an efficient process would resolve the dispute in a timely, cost-effective manner. One element of an efficient process involves data sharing. With the current process, the two national sections conduct independent investigations and then share the data. However, no provisions call for sharing data with a broader spectrum of stakeholders. Although gages maintained by the IBWC provide easily accessible data about streamflows and reservoir water volumes, little data sharing takes place between the two sections regarding water use issues. The lack of publicly available data resulted in major disagreements among local stakeholders on both sides of the border over how much water was available to users and served to intensify conflicts over water deliveries at the local and state levels in both Texas and Mexico.

A specific example where data sharing could have played a role in attenuating transboundary conflicts is open sharing of water use information between the two countries. Irrigation water allotments in the Delicacias Irrigation District in the Rio Conchos watershed decreased 31% over the period 1993–2000, falling below those during the period prior to 1992.[140] In all years but 1997, no water releases were made for winter crops. Water storage in Mexican reservoirs in 2000 was about 26% for the Conchos basin and 11% for the Salado. Farmers in the Lower Rio Grande Valley in Texas began receiving reduced water deliveries from the international storage reservoirs for irrigation.[141] Both sides believed the other had an advantage, because neither had access to the other country's data on reservoir conditions and usage. Texas farmers believed that Mexican farmers were using water in the Rio Conchos that should have passed to Texas pursuant to the terms of the 1944 Treaty. Texas interests asserted, among other things, that Mexican reservoirs were full and irrigation in the Rio Conchos basin was increasing. Mexican farmers in Tamaulipas, downstream on the Rio Grande, believed that water transferred to the United States came at the expense of their allotments for irrigation.

The disagreement over who was using whose water led to increasingly acrimonious statements from local politicians on both sides of the border. The governor of Texas threatened an end to diplomatic efforts, while the Mexican Congress passed a resolution stating that no water deliveries would be made because Mexican water uses came first.[142] Texas's commissioner of agriculture handed Mexican president Vicente Fox satellite imagery purportedly showing water in Mexican reservoirs that Mexico could pass downstream to Texas.[143] Lack of discussion between the two sides and the failure to share information from the outset of the dispute exacerbated the conflict. Further, no forum existed to allow for input from local citizens or environmental groups concerning factual information.

Minute 308 indicates the intent of the IBWC to foster data sharing between the two countries, but it retains the current system whereby the parties work separately and forward their data and reports to each other in a "timely manner."[144] Data sharing certainly is a step forward, but this minute does not include any objective criteria for joint decisionmaking. It merely states that Mexico will provide a progress report on drought planning to the commission, in order to support the commission as a forum under which proper authorities in each country can coordinate drought management.[145] Although Minute 308 is evidence of the desire of the two nations to enhance data sharing, the structure of the agreement perpetuates the top-down decisionmaking style that appears to have contributed to problems at the state and local levels during the water dispute.

The second measure of evaluating negotiated agreements involves assessment of the fairness of the dispute resolution process. One obvious attribute of a fair process would be a forum allowing all interests to be heard, including accountability to the public. Interests that should be at the table might include water users, nongovernmental organizations, and representatives of the public at large. In the case of the 1944 Treaty process, few avenues exist for water users to participate directly in negotiations, except perhaps for communicating their concerns to their section of the IBWC. Evidence of the futility of this method of inclusion is found in the rhetoric employed by state-level politicians, answerable to local interests, in a somewhat futile effort to influence negotiations over the Mexican water debt. Because the formal negotiation process has no mechanism to include stakeholder concerns, the communication of local stakeholder interests can degenerate into name-calling and threats. With Texas interests on water allocation issues subordinate to broader U.S. interests, as evidenced by the U.S. filing in the NAFTA case, and with water issues considered less important than other issues, such as illegal immigration, to the foreign policy of the United States, the reaction of Texans was understandable. The only forum in which to air concerns was the local media.

To add to issues regarding stakeholder representation, there are no established procedures for inclusion of the interests of environmental groups. Environmental groups from both nations convened their own summit to discuss sustainable basin management and drought planning. The groups issued a binational declaration calling for increased water use efficiency, drought planning, and inclusion of environmental flows as part of any plan for sustainable management of the basin.[146]

However, one sector's interests appear to be adequately represented in the current conflict resolution process. That sector is federal governments on both sides of the border, through the IBWC and the two countries' State Departments. The interests of the United States and Mexico, insofar as those interests are limited to maintain a cordial relationship, are well served by the current process. The various minutes issued during the conflict attest to continued friendship, and there is a definite tendency to state overarching interests, such as the fundamental human right to water for basic uses. Thus, the process appears to be fair, at least at the international and national levels, in the sense that the national governments are able to agree, maintain cordial relationships, and set aside differences in the broader public interest.

The final measure of a negotiated agreement is durability. Negotiated agreements should set the stage for a comprehensive solution, be equitable to both current and future generations, and allow for continuing dialogue among the parties as circumstances change. All relevant parties should be included to ensure fairness of the process. This is necessary so that all issues are addressed during the negotiations, and it allows for sufficient buy-in from every stakeholder in order to facilitate implementation of any agreement. The USIBWC has instituted citizens' forums to aid in the exchange of information about its activities in all border regions, including both the Lower and Upper Rio Grande.[147] Although this is a step in the right direction and points to efforts by the USIBWC to make the process more transparent, the process itself is not formalized to the extent that stakeholder concerns are required to be considered in the operations of this institution. Note that a comprehensive solution does not mean that all problems are immediately solved in a single negotiation. Rather, it means that the door is open so that a comprehensive agreement can be reached over time through further negotiation. Thus, a comprehensive agreement sets the stage for a continuing dialogue, and a durable agreement should contain provisions for intergenerational equity.

Minutes 307 and 308 do allow for both continuing dialogue and addressing intergenerational equity, in the guise of initiating discussion of sustainable water use in the Rio Grande basin. The minutes did not define either sustainable water use or sustainability in general, and they did not establish concrete processes to achieve these goals. The basinwide summit, agreed on in 2002, occurred in November 2005.[148] The conference was organized as a series of presentations followed by issue-oriented work groups on such topics as finance, environment and water quality, and legal and institutional issues. The work group recommendations were not available until August 2007. Although organization of the summit was a step in the right direction, lack of an overarching framework for incorporation of recommendations, timeliness of issues related to information availability, or a concrete public plan for moving forward tend to blunt the effectiveness of this mechanism.

An additional shortcoming of Minute 308 is the two governments' tepid endorsement of the suggestion to establish an International Advisory Council. By failing to include a mechanism to ensure that all stakeholders interests are accounted for in the negotiation process, any negotiated settlements generated to try to address issues resulting from the latest conflict, as well as conflicts over other issues that may occur in the future, likely would fail to meet the standard of comprehensiveness. This would tend to reduce the probability that any negotiated agreement would be durable in the long term.

Post-Treaty Administration: Water Quality Issues

The 1944 Treaty also grants the IBWC authority to address water quality issues.[149] The first real efforts of the IBWC to address water quality occurred in the late 1960s, when a salinity crisis occurred involving the Colorado River. Although the United States complied with treaty requirements related to quantity, the water was unusable because of elevated salinity attributed to U.S. agricultural return flows. The issue was resolved via negotiations, resulting in Minute 242, which established a salinity requirement for treaty water deliveries.

Minute 261, signed in 1979, began the process of addressing border sanitation issues. Issues related to water quality also can be contentious, as they frequently involve competing interest groups and differing regulatory regimes in both countries.[150] The United States signed an agreement in 1983 to address border environmental issues, known as the La Paz Agreement.[151] A direct result of this agreement was Border 2012, a program jointly managed by the U.S. Environmental Protection Agency (EPA) and Mexican Secretaria de Medio Ambiente y Recursos Naturales (SEMARNAT). Border 2012 uses a bottom-up approach to address environmental and public health issues in the U.S.-Mexico border region.[152] Furthering progress in the area of water pollution, through establishment of the Border Environmental Cooperation Commission (BECC) and the North American Development Bank (NADBank), NAFTA provides venues for project proposals and financing. Minute 279 provided for an international wastewater treatment plant at Nuevo Laredo. In Minute 289, the United States and Mexico began investigations related to the presence of toxic substances in the Rio Grande.

CONCLUSIONS

Water allocation frameworks designed for past conditions may not be adequate for twenty-first-century problems. Although Texas's post-compact relationships with its neighbors could be characterized as litigious, interstate compacts do provide a level of certainty with respect to water allocations. Indeed, the threat of litigation, including the costs associated with resorting to this alternative, often acts to keep states participating at the negotiating table. In addition to cost factors related to litigation, the court process also is time-consuming, contributing to policy lags, which may not be in the best interest of local Texas water users subject to the compacts.

Compacts in geographic areas more subject to the vagaries of climate, particularly drought, typically experience more issues related to enforcement of compact requirements. Relevant examples include the Rio Grande, Pecos, and Canadian basins. Although endangered species issues currently are being handled within the current structure of the compacts, the next severe drought could propel administration of these agreements back into the legal realm. Further, water marketing across interstate boundaries is a developing issue, as thirsty Texan cities attempt to procure water across state lines.

With respect to Texas's relationships with its international neighbor, some aspects are working relatively well, an example being the management of water quality problems. Reluctance to define extreme drought and to incorporate mechanisms for joint drought management inevitably will return Texas water users to the same state of affairs that existed for 10 years during the previous drought. Some progress has been made, such as binational initiatives addressing sustainable use of the Rio Grande. Lack of a comprehensive framework and process to address this issue, however, may hamper such efforts. Efforts between the United States and Mexico to undertake a transboundary diagnostic analysis and implement a strategic action plan for the Rio Grande basin, currently under way through the auspices of the Global Environment Facility and United Nations Environmental Program, may assist in this regard.

Two problem areas the region is most likely to have to deal with in the future are binational groundwater issues and environmental flows. Groundwater issues in the Rio Grande region, even for transboundary aquifers, currently are addressed by each state. The role of the USIBWC to date has been to gather information on these shared water resources. If negotiations of surface-water allocation are any indication, mechanisms for shared aquifer management face hurdles, in view of current Texas groundwater law and the extended time frames required to negotiate agreements. Incorporating environmental flows into the treaty framework also is likely to be many years in the future, although Texas's Instream Flow Programs may provide information useful in such an endeavor.

Climate change may have unforeseen impacts in the Rio Grande basin, particularly because current water allocation and supply management regimes are governed by compacts and treaties that require delivery of specific volumes of water to downstream states. This raises the question of whether the existing legal framework is sufficiently flexible to address the possible effects of climate change. These effects may include variations in existing precipitation patterns; decreased snowmelt in Colorado; earlier snowmelt; increased winter rains, which could severely affect the seasonality of the river's flow regime; or increased temperatures, which may have a drastic impact on water supplies by increasing the volume of water lost to evaporation. In addition to exacerbating water scarcity, climate change also could cause increased flooding. This could be a major issue in low-lying areas in the Rio Grande Valley on both sides of the border.

International law, as reflected in the 1997 United Nations Convention on the Non-Navigational Uses of International Watercourses, is relatively undefined in the area of water reallocation resulting from climate change. Further, the 1944 Treaty and the river compacts are fairly inflexible with respect to water reallocation issues, containing no provisions that allow adaptation in the event of altered circumstances such as climate change, water to protect instream uses and aquatic species, or population growth. Moving forward on any of these issues is a major challenge for the future of the Rio Grande basin.

NOTES

1. For a detailed discussion of general legal issues related to compacts, see Priscilla Hubenak and Tom Bohl, Multi-jurisdictional Water Rights, in *Essentials of Texas Water Resources*, ed. Mary K. Sahs, 235 (Austin, TX: State Bar of Texas Environmental and Natural Resources Law Section, 2009).

2. Texas State Historical Association, Handbook of Texas Online, www.tshaonline.org/handbook/online/articles/CC/mgc1.html (accessed June 1, 2010).

3. U.S. Bureau of Reclamation, http://www.usbr.gov/projects/Project.jsp?proj_Name=Tucumcari Project&pageType=ProjectPage (accessed June 4, 2010).

4. P.L.81-898, Canadian River Reclamation Project, Texas, Senate Report No. 81-2110, July 20, 1950 (to accompany HR 2733).

5. Paul Elliott, Texas's Interstate Compacts, *St. Mary's Law Journal* 17 (1986): 1261.

6. Act of April 29, 1950, ch. 135, P.L. 81-491, 64 Stat 49.

7. Act of May 17, 1952, ch. 306, P.L. 82-345, 66 Stat. 74.

8. Texas Water Code Ann, § 43.006, art. II(d), (West 2008).

9. Ibid., art. IV(a), (b).

10. Ibid., art. VII.

11. Ibid., art. V(a).

12. Ibid., art. V(b).

13. Elliott, Texas's Interstate Compacts, 1262.

14. *Oklahoma v. New Mexico*, 501 U.S. 221, 111 S. Ct. 2281, 226–27.

15. Ibid.

16. Ibid., 228.

17. *Oklahoma v. New Mexico*, 510 U.S. 126, 114 S. Ct. 628, 629.

18. Oklahoma Water Resources Board, Resolution Asks Support in Compact Dispute, Board Convinced All Other Remedies Are Exhausted in Securing Oklahoma's Share of Palo Duro Water, *Oklahoma Water News*, May–June 2000, http://www.owrb.ok.gov/news/news2/pdf_news2/WaterNews/WaterNews2000-3.pdf (accessed June 4, 2010).

19. Hubenak and Bohl, Multi-jurisdictional Water Rights.

20. Handbook of Texas Online, s.v. "Red River," available at http://www.tshaonline.org/handbook/online/articles/RR/rnr1.html (accessed June 4, 2010).

21. Elliott, Texas's Interstate Compacts, at 1273; Pub. L. 96-584, 94 STAT. 3305, December 22, 1980.

22. Texas, Ann § 46.013, art. XIII, § 13.03, (West 2008).

23. Ibid., art. XI.

24. Ibid., art. II, § 2.11.

25. Red River Compact Commission, Agency Strategic Plan for the Fiscal Years 2005–2009 Period, 2005, http://www.tceq.state.tx.us/assets/public/permitting/watersupply/water_rights/splan.red.pdf (accessed June 4, 2010).

26. Texas Water Code Ann, § 46.013, art. IV, §§ 4.01–4.04, (West 2008).

27. Ibid., art. V, §§ 5.01–5.05.

28. Ibid., art. VI, §§ 6.01–6.03.

29. Red River Compact Commission, *Resolution to Adopt Rules and Regulations to Compute and Enforce Compact Compliance Reach I, Subbasin 1-Sweetwater Creek and North Fork Red River*, adopted April 22, 2008 at Marshall, TX (on file with author).

30. Jim Williamson, Texas Goes Shopping for Water, *Texarkana Gazette.* January 23, 2007.

31. *Tarrant Regional Water District v. Rudolf Herrmann et al.*, WL 3226812, W.D. Okla., 2007; *Tarrant Regional Water District v. Sevenoaks*, F. 3d. WL 4694903, C.A. 10 (Okla.), October 27, 2008.

32. Max Baker, Irving Criticized as It Plots Its Own Course to Get Water from Oklahoma, *Fort Worth Star-Telegram*, August 10, 2008.

33. Rudolph Bush, Lawsuit Ruling Edges Tarrant, Dallas Districts Closer to Oklahoma Water Supply, *Dallas Morning News*, October 28, 2008.

34. Elliott, Texas's Interstate Compacts, 1263.

35. Texas Water Code Ann, § 44.010 at art. III, (West 2008).

36. Ibid., art. V(a).

37. Ibid., art. V(i).

38. Ibid., art. V(e).

39. Ibid., art. V(b).

40. Sabine River Authority of Texas, Toledo Bend Project, http://www.sratx.org/projects/tbp.asp (accessed June 4, 2010).

41. Vickie Welborn, Texas Water Meetings Could Impact Toledo Bend, *Shreveport Times*, May 5, 2008.

42. David J. Eaton and John M Andersen, *The State of the Rio Grande/RioBravo: A Study of Water Resource Issues along the Texas/Mexico Border* (Tucson, AZ: University of Arizona Press, 1987).

43. Norris Hundley, Jr, *Dividing the Waters: A Century of Controversy between the United States and Mexico* (Berkeley, CA: University of California Press, 1966), 19.

44. Jurgen Schmandt, *Water and Development: The Rio Grande/Rio Bravo* (Austin, TX: Lyndon B. Johnson School of Public Affairs, 1993).

45. Texas Water Development Board (TWDB), Region M Water Plan, 2006, http://www.twdb.state.tx.us/RWPG/main-docs/2006RWPindex.asp (accessed June 4, 2010).

46. Jurgen Schmandt, Ismael Aguilar-Barajas, Neal Armstrong, Liliana Chapas-Aleman, Salvador Contreras-Balderas, Robert Edwards, Jared Hazleton, Jose de Jesus Navar, Mitchell Mathis, Enrique Vogel-Martinez, and George Ward, *Final Report: Water and Sustainable Development in the Binational Lower Rio Grande/Bravo Basin*, EPA Grant No. R824799 (The Woodlands, TX: Houston Advanced Research Center, Center for Global Studies, 2000).

47. William Paddock, The Rio Grande Convention of 1906: A Brief History of an International and Interstate Apportionment of the Rio Grande, *Denver University Law Review* 77 (1999): 287, 295–296, 297, 300.

48. Act of June 17, 1902, 57th cong. (1st sess.) ch. 1093, 32 Stat. 388, codified at 43 U.S.C. § 371.

49. U.S. Bureau of Reclamation, Rio Grande Project: History, http://www.usbr.gov/projects/Project.jsp?proj_Name=Rio Grande Project&pageType=ProjectHistoryPage (accessed June 4, 2010).

50. Paddock, Rio Grande Convention of 1906, 305.

51. Convention for the Equitable Distribution of the Waters of the Rio Grande for Irrigation Purposes, U.S.-Mexico, May 21, 1906, 34 Stat. 2953, available at http://www.ibwc.state.gov/Files/1906Conv.pdf (accessed June 4, 2010).

52. Ibid., art. I.

53. Ibid., arts. III, IV.

54. Handbook of Texas Online, s.v. "Pecos River," available at http://www.tshaonline.org/handbook/online/articles/PP/rnp2.html (accessed June 4, 2010).

55. G. Emlen Hall, *High and Dry: The Texas–New Mexico Struggle for the Pecos River* (Albuquerque, NM: University of New Mexico Press, 2002).

56. Ric Jensen, Will Hatler, Mike Mecke, and Charlie Hart, The Influences of Human Activities on the Waters of the Pecos Basin of Texas: A Brief Overview, 2005, available at http://twri.tamu.edu/reports/2006/sr2006-03.pdf (accessed June 4, 2010).

57. Elliott, Texas's Interstate Compacts, 1252.

58. Ibid., 1253.

59. Robert Lingle and Dee Linford, *The Pecos River Commission of New Mexico and Texas, A Report of a Decade of Progress 1950–1960* (Santa Fe, NM: Rydal Press, 1961), 125, 134, 136–38.

60. Texas Water Code Ann, § 42.010, (West 2008).

61. Hubenak and Bohl, Multi-jurisdictional Water Rights.

62. Elliott, Texas's Interstate Compacts, 1255.

63. *Texas v. New Mexico*, 462 U.S. 554, 103 S. Ct. 2258 at 2563 (1983).

64. Ibid., 564; Hubenak and Bohl, Multi-jurisdictional Water Rights.

65. *Texas v. New Mexico*, 494 U.S. 111, 110 S. Ct. 1293.

66. Pecos River Compact Commission, Texas, *Agency Strategic Plan for the 2005–2009 Period*, 2005, available at http://www.tceq.state.tx.us/assets/public/permitting/watersupply/water_rights/splan.pecos.pdf (accessed June 4, 2010)

67. Douglas L. Grant, Interstate Water Allocation Compacts: When the Virtue of Permanence Becomes the Vice of Inflexibility, *University of Colorado Law Review* 74 (2003): 111–12.

68. Pecos River Compact Commission, *Strategic Plan for 2005–2009*.

69. Douglas Littlefield, The Rio Grande Compact of 1929: A Truce in an Interstate River War. *Pacific Historical Review* 60, no. 4 (1991): 497–515.

70. Ibid., 503.

71. Ibid., 504, 508, 513.

72. Ibid., 514.

73. William Paddock, The Rio Grande Compact of 1938, *University of Denver Water Law Review* 5, no. 1 (2001): 17.

74. For a detailed description of the negotiations, see ibid., 18–27.

75. Texas Water Code Ann, § 41.009, (West 2008).

76. Elliott, Texas's Interstate Compacts, 1250.

77. Paddock, Rio Grande Compact of 1938, 42–44.

78. *Federal Register* 36988. Final rule listing the Rio Grande silvery minnow as an endangered species.

79. Grant, Interstate Water Allocation Compacts, 105; Rio Grande Compact Commission, Strategic Plan and Compact with Texans, 2004, www.tceq.state.tx.us/assets/public/permitting/watersupply/water_rights/splan.rio.pdf (accessed June 4, 2010).

80. Hundley, *Dividing the Waters*, 30–31.

81. Ibid., 39–40.

82. Ibid., 76–77.

83. Charles A. Timm, *The International Boundary Commission: United States and Mexico*, Publication No. 4134 (Austin, TX: University of Texas, 1941).

84. Ibid.

85. Hundley, *Dividing the Waters*, 93–94.

86. Ibid., 41–42.

87. Timm, *International Boundary Commission*.

88. Hundley, *Dividing the Waters*, 94.

89. Timm, *International Boundary Commission*.

90. Hundley, *Dividing the Waters*, 94.

91. Timm, *International Boundary Commission*.

92. Treaty between the United States of America and Mexico, February 3, 1944, U.S.-Mex. 59 Stat. 1219 (referred to in the text as the 1944 Treaty), art 4.

93. Ibid., art. 3.

94. Ibid., art. 4.

95. Ibid., art. 5.

96. Ibid., art. 2.

97. Ibid., art. 24.

98. Ibid., art. 25.

99. Ibid., art. 4.

100. International Boundary and Water Commission (IBWC), *Commission Report, United States and Mexico* (El Paso, TX: IBWC, 1998).

101. Treaty, arts. 10–16.

102. Hundley, *Dividing the Waters*, 93–97.

103. John Haywood, *A Brief Legal History of the Irrigation Rights Controversy of Land Originating in Spanish and Mexican Grants in the Rio Grande Valley of Texas* (n.p., 1974).

104. *Motl v. Boyd*, 116 Texas 82, 286 S.W. 458.

105. *State of Texas v. Valmont Plantations*, 163 Tex. 381, 355 S.W.2d 502.

106. Matthews, Nowlin, MacFarlane, & Barrett, *Water Rights in the Lower Rio Grande Valley: An Outline of the Principle Issues of Law and Fact in the Case of the State of Texas vs Hidalgo County Water Control and Improvement District No. 16, et al. in the District Court of Hidalgo County, Texas*, 1957.

107. 1944 Treaty, art. 5.

108. Matthews, Nowlin, MacFarlane, & Barrett, *Water Rights*.

109. Ibid.

110. *State of Texas v. Hidalgo County*, WCID No. 18, 443 S.W.2d 7281 (1969).

111. Jurgen Schmandt, Chandler Stolp, and George H. Ward, *Scarce Water: Doing More with Less in the Rio Grande*, Report for the Binational Assessment of Water and Development in the Lower Rio Grande/Rio Bravo Basin, U.S.-Mexican Policy Report No. 8 (Austin, TX: Lyndon B. Johnson School of Public Affairs, 1988).

112. Texas Water Code Ann, § 11.326, (West 2008).

113. Texas Administrative Code, § 303.21(b)(1).

114. Ibid., § 303.22(c).

115. Texas Water Code Ann, § 11.122, (West 2008).

116. Texas Administrative Code, § 303.53.

117. Rio Grande Regional Water Planning Group, Rio Grande Regional Water Plan, January 5, 2005, available at http://www.riograndewaterplan.org (accessed June 4, 2010).

118. Damien Schiff, Rollin', Rollin', Rollin' on the River: A Story of Drought, Treaty Interpretation, and Other Rio Grande Problems, *Indiana International and Comparative Law Review* 14 (2003): 26.

119. Texas Center for Policy Studies, The Dispute over the Shared Resources of the Rio Grande/Rio Bravo: A Primer, 2002, available at http://www.texascenter.org/borderwater (accessed June 4, 2010).

120. 1944 Treaty, art. 4.

121. House Research Organization, Behind the U.S.-Mexico Water Treaty Dispute, *Interim News* 77-7 (April 30, 2002). For a detailed accounting of deliveries and deficits, see Joe G. Moore, Jr, Walter Rast, and Warren Pulich, Proposal for an Integrated Management Plan for the Rio Grande/Rio Bravo, in *First International Symposium on Transboundary Waters Management, Avances en Hidraulica 10, XVII Mexican Hydraulics Congress*, ed. A. Aldama, F. Aparicio, and R. Equihua, 189–204 (Monterrey, Mexico: November 18–22, 2002).

122. Texas Center for Policy Studies 2002, 11.

123. 1944 Treaty, min. 307.

124. Texas Center for Policy Studies 2002, at 13.

125. 1944 Treaty, min. 307, #3.

126. IBWC, *Report of the United States Section International Boundary and Water Commission Deliveries of Waters Allotted to the United States under Article 4 of the United States Mexican Water Treaty* (El Paso, TX: IBWC, 2002).

127. 1944 Treaty, min. 308.

128. Ibid., rec. 5.

129. Ibid., part G.3.

130. Ibid., rec. 5.

131. Stephen P. Mumme, Revising the 1944 Water Treaty: Reflections on the Rio Grande Drought Crises and Other Matters, *Journal of the Southwest* 45, no. 4 (2003): 649–70.

132. Paul Krza, Texas Water Case Is "Takings" on Steroids, *High Country News*, February 21, 2005.

133. *Bayview Irrigation District et al. v. United Mexican States*, ICSID Case No. ARB(AF)/05/01, before the Arbitral Tribunal Constituted under Chapter XI of the North American Free Trade Agreement, Award June 19, 2007.

134. Luke E. Peterson, Texas Farmers Sue Mexico in Ontario for Water, *Embassy*, November 17, 2007.

135. *Bayview et al. v. United Mexican States*.

136. Ibid.

137. *Bayview Irrigation District No. 11 and ors v. Mexico*, Judicial Review, 07-CV-340139-PD2;IIC 323 (2008), May 5, 2008.

138. 1944 Treaty, min. 308, part A.

139. Barbara Cosens, Water Dispute Resolution in the West: Process Elements for the Modern Era in Basin-Wide Problem Solving, *Environmental Law* 33 (2003): 949.

140. Texas Center for Policy Studies 2002, 8.

141. Ibid., 11.

142. Ibid., 14.

143. South Texans Take Water Fight with Mexico to Canada Court, *Dallas Morning News*, February 19, 2008.

144. 1944 Treaty, min. 308, rec. 4.

145. Ibid., part G.

146. Environmental Defense Fund, Discovering the Rio Conchos, May 3, 2002, http://www.edf.org/article.cfm?ContentID=2915 (accessed June 4, 2010).

147. U.S. Section of the International Boundary and Water Commission (USIBWC). For available information regarding meeting minutes and presentations, see USIBWC, Lower Rio Grande Citizens' Forum Meetings, http://www.ibwc.gov/Citizens_Forums/CF_Lower_RG.html (accessed June 4, 2010).

148. Conference proceedings and recommendations are available at USIBWC, Binational Rio Grande Summit Recommendations, http://www.ibwc.gov/Organization/rg_summit_recommendations.html (accessed June 4, 2010).

149. 1944 Treaty, art. 3.

150. William Wilcox, Mexico and the United States: The Trans-boundary Water Quality Issues That Lie Ahead, *Western Water Law & Policy Reporter* (April 1999): 137–41.

151. Agreement on Cooperation for Protection and Improvement of the Environment in the Border Area, August 14, 1983, United States–Mexico, 22 I.L.M. 1025.

152. Specific information on Border 2012 can be found at U.S. EPA, U.S.-Mexico Border 2012 Program, http://www.epa.gov/usmexicoborder/index.html (accessed June 4, 2010).

CHAPTER 9

Groundwater Depletion in the Texas High Plains

David B. Willis and Jeffrey W. Johnson

*I*n the Texas High Plains (THP), the Panhandle region of northwest Texas, groundwater for agriculture accounts for 96% of all Southern Ogallala Aquifer withdrawals.[1] The heavy agricultural reliance on groundwater is attributable to limited surface supplies and the high cost of developing surface-water storage within the region. Annual withdrawals from the Southern Ogallala Aquifer have been estimated to be more than 10 times greater than the natural recharge rate,[2] so the aquifer is being mined as an exhaustible resource.

THP agriculture has developed alongside Texas groundwater law, commonly referred to as the rule of capture. This law grants landowners the right to all groundwater that may be accessed from their land (see Chapter 3), and irrigators have exploited this privilege to tap vast Southern Ogallala Aquifer groundwater reserves that had been untouched for thousands of years. As a result of heavy agricultural withdrawals over the last 50 years, Southern Ogallala groundwater reserves are at approximately 50% of their 1940 storage level.[3]

Recent Texas legislation (Senate Bills 1 and 2) explicitly recognizes the growing scarcity of the state's groundwater supplies and requires the Texas Water Development Board (TWDB) to develop a statewide water use plan that incorporates locally developed regional water plans. When Senate Bill (SB) 1 was passed in 1997, it modified several sections of the Texas Water Code and required the coordinated development of a comprehensive statewide water management plan that incorporated locally developed regional water plans.[4] SB 1 identified the TWDB as the lead state agency in the statewide regional planning effort, and the TWDB subsequently defined 16 regional water-planning groups. Region O encompasses 21 Texas counties overlying the Southern Ogallala Aquifer, and Region A consists of 21 counties in the northern Panhandle region. Each regional planning group was required to develop a 50-year management plan.

Enacted by the 77th Texas legislature in 2001, SB 2 complements the earlier bill by supporting statewide cooperation among various state agencies and local planning groups.[5] A weakness in the original SB 1 legislation was that it lacked the formal institutional infrastructure needed to coordinate statewide water use planning. SB 2 addressed this omission by establishing the Texas Water Advisory Council and additional management guidelines.[6] It increased the authority of groundwater districts to implement production fees or taxes on agricultural withdrawals for groundwater management purposes, but the fees are limited to a maximum value of $1 per acre-foot (ac-ft) of water withdrawn. This bill also instructed the TWDB to identify groundwater management areas (GMAs) sufficiently large to accommodate hydrologic boundaries, and to facilitate joint planning among groundwater districts and municipal jurisdictions sharing the same groundwater resource.

In 2002, the TWDB identified 16 GMAs, the boundaries of which generally differ from the previously established regional planning group boundaries, because they now reflect both politico-economic and hydrologic linkages, whereas counties formerly were grouped together according to shared politico-economic linkages. It is anticipated that each GMA will facilitate the development of a sustainable 50-year water management plan that equitably treats the economic and social interests of the managed area.[7] GMA 2 is the Southern Ogallala Aquifer management area of the THP, and GMA 1 covers the Ogallala Aquifer in the northern Panhandle.

Collectively, SB 1 and SB 2 are steps in the transition of Texas groundwater management from the rule of capture, which is a private property doctrine of land (not water; see Chapter 3) providing groundwater access via landownership, to a regulatory system that enables local and state government agencies to monitor and regulate the volume of groundwater extracted.[8] How this system will evolve beyond these steps is presently unknown.

This chapter analyzes the effectiveness of the available groundwater conservation tools to extend the economic and physical life of the Southern Ogallala Aquifer. As a prelude to the analysis, an overview of the Ogallala Aquifer system is provided, followed by a brief history of irrigation development in the THP, an overview of the THP region's economy, and the legal and institutional issues affecting groundwater management there. The chapter concludes with an analysis of the conservation effectiveness of current groundwater policy tools and a discussion of the policy implications and future research needs. The THP water policy model developed by Das and Willis is used to estimate the volume of groundwater conserved and the acre-foot cost of water conservation imposed on irrigated agricultural producers, the dominant water user in the region, over a 50-year planning horizon.[9] Findings reveal that over the next 50 years, the currently available conservation policy tools, if fully implemented, will decrease agricultural groundwater withdrawals in the THP by less than 3% of the expected withdrawal level.

THE OGALLALA AQUIFER SYSTEM

The Ogallala Aquifer is one of the largest aquifer systems in the world, encompassing an area of 174,000 square miles and underlying portions of eight

Great Plains states: Texas, New Mexico, Colorado, Oklahoma, Kansas, Nebraska, Wyoming, and South Dakota.[10] The Canadian River Valley and the Prairie Dog Fork of the Red River Valley divide the Southern High Plains and Central High Plains regions of the Ogallala Aquifer.[11] The northern border of the Southern Ogallala is located in the Texas Panhandle and is minimally connected to the Central Ogallala, which is primarily located in Kansas and Oklahoma but extends southward into the northern area of the Texas Panhandle. As illustrated in Figure 9.1, the 42,000-square-mile Southern Ogallala Aquifer comprises the southernmost quarter of the Ogallala Aquifer system and is located entirely in Texas (85%) and New Mexico (15%).[12] There is very little hydraulic connectivity between the Southern and Central Ogallala Aquifers.[13]

The Southern Ogallala Formation ranges in depth, below ground surface, from 0 to 800 feet in the north near the Canadian River to 500 feet in the south near Midland County. Recharge into the Southern Ogallala Aquifer occurs through 25,000 playa lakes and numerous riverbeds located within the region, as well as

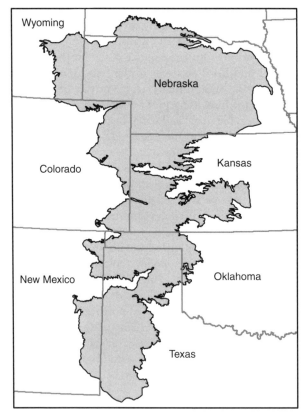

Figure 9.1 *The Ogallala Aquifer system*

Source: Center for Geospatial Technology, *Impact of Alternative Water Policies on Rural Economies and Aquifer Hydrology: Development of a GIS Database of Hydrologic Characteristics for the Ogallala Aquifer Program* (Lubbock, TX: Texas Tech University, 2008).

diffuse infiltration into Ogallala outcroppings on the western and northern boundaries. Nativ determined that groundwater infiltration through riverbeds is minimal because of poorly developed drainage systems. The caliche layer, known regionally as the caprock, lies above the formation and has a low permeability, which restricts diffuse recharge. Thus, most recharge occurs through the region's playa lakes. Average annual recharge rates for the Southern Ogallala Aquifer range between 0.11 and 1.60 inches.[14]

HISTORY OF IRRIGATION DEVELOPMENT

Agricultural settlements began to dot the THP landscape in the late 1870s. Groundwater wells were first hand dug near present-day Lubbock in 1878. Early wells were powered by horses and used an endless chain and bucket system to withdraw groundwater. In areas conducive to windmill power, this technology quickly replaced horse-driven power.[15] Windmills were inexpensive and provided settlers with easy water access in a region with limited surface supplies, but they could not dependably support large irrigated farms in prolonged drought periods, nor could they pump water from depths greater than 70 feet.

The centrifugal pump, the first non-wind-powered pump, was introduced in 1900. It could lift several hundred gallons per minute but could not lift water more than 20 feet, which restricted pump adoption to areas with low pump lifts.[16] In areas where the water table was more than 20 feet below the surface, a pit needed to be dug near the well so that the pump could be placed within reach of the water table. The "pitless" pump, developed in 1903, overcame the need to dig a pit to lift groundwater more than 20 feet, required less maintenance, and was powered by a low-cost, low-maintenance oil-burning engine.[17] Despite advances in pumping technology, irrigated acreage did not significantly increase during 1910–1920. Above-average rainfall levels minimized the need to invest scarce capital in irrigation technology, and higher wheat and livestock prices driven by World War I demands favored more extensive dryland cropping rather than smaller intensive irrigation cropping practices in this decade.[18] Chronically depressed crop prices throughout the 1920s, resulting from the large increase in crop production levels associated with the war effort of the prior decade, hindered the growth of irrigated acreage throughout the 1920s.

Irrigated acreage expanded rapidly in the 1930s with the return of devastating droughts and the decreased cost of installing groundwater pumps. The number of irrigation pumping plants increased from 170 in 1930 to 2,680 in 1940, when 250,000 acres were irrigated.[19] As irrigation became more commonplace, crop rotations began to switch from grains and forage crops for livestock production to greater specialization in cotton and wheat.

World War II demands and price supports for cotton and wheat, along with intensive crop production in the region, drove additional increases in irrigated acreage throughout the 1940s and 1950s. The growing availability of electricity and natural gas complemented the adoption of irrigated production practices. These new fuel sources allowed wells to be profitably installed in areas of the THP with high groundwater lifts, fueling an increase in the number of irrigation wells from 2,560 in 1939 to 44,630 in 1958. Irrigated acreage correspondingly grew from 200,065 to 4,535,941 during this time interval, as illustrated in Figure 9.2.

By 1960, production agriculture in the THP had transitioned from small subsistence farms to large, profitable farms with substantial investments in irrigation technology. Moreover, irrigation technology was being adopted to increase crop yield and profitability rather than as a form of self-insurance against drought. Irrigation technology had transformed the once drought-ridden THP into a very productive agricultural region and spurred population growth here when many other west Texas counties east and south of the irrigated THP were beginning to lose population.[20] However, this economic development came at a cost. By the mid-1970s, three decades of heavy groundwater withdrawals had lowered the water table more than 200 feet from predevelopment levels in some areas.[21] Increased pump lifts associated with the dropping water table in combination with high energy prices led to a contraction in irrigated acreage, which dropped from a high of nearly 6 million acres in 1974 to slightly less than 4 million acres in 1989 (see Figure 9.2).

Low energy precision application (LEPA) irrigation systems were developed in the late 1980s to reduce the costs of energy and increasing pump lift. Water application efficiency with LEPA systems is greater than with the previously used center pivot sprinkler and furrow irrigation systems.[22] An unintended consequence of improved irrigation efficiency was that improved application efficiency lowered the per-acre cost of applying the quantity of water sufficient to

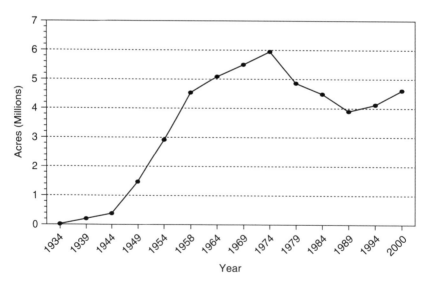

Figure 9.2 *Irrigated crop acreage in the Texas High Plains, 1934–2000*

Sources: United States Census of Agriculture: 1935 (Table VI), 1940 (Table I), 1945 (Table I), 1950 (Table I), and 1955 (Table Ia); Texas Water Development Board (TWDB), *Surveys of Irrigation in Texas*, Report 347 (Austin, TX: TWDB, August 2001): 1958, 1964, 1969, 1974, 1979, 1984, 1989, 1994, and 2000 data.

satisfy crop net irrigation requirements, and irrigated acreage began to increase again—from 3.9 million to 4.6 million between 1989 and 2000, as shown in Figure 9.2—even though crop and energy prices remained relatively stable. Despite the introduction of more water-efficient LEPA irrigation systems, the water table continued to decline throughout the 1990s in many areas of the THP.[23] Irrigation technology improvements in combination with crop yield improvements throughout the latter part of the twentieth century allowed irrigated agriculture to remain a profitable economic activity in the region despite the continued drawdown of the Southern Ogallala Aquifer.

THE TEXAS SOUTHERN HIGH PLAINS REGIONAL ECONOMY

The Texas portion of the Southern Ogallala Aquifer is bounded in the west by the New Mexico state line, to the south and east by the Caprock Escarpment, and by the Canadian River in the northern Panhandle region. The gray shaded area in Figure 9.3 identifies the 19 Texas counties that account for 97% of all agricultural groundwater diversions from the Texas portion of Southern Ogallala Aquifer.

The region is largely rural and has a population exceeding 500,000, but only five of the area's cities have populations greater than 10,000.[24] Typical of many rural areas, the regional economy is dominated by agriculture, with production agriculture accounting for 25% of all employment in 2002. Cotton and beef production are the two dominant agricultural activities.[25] One-third of all personal

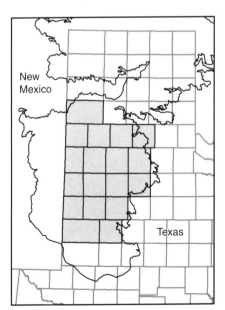

Figure 9.3 *The Southern Ogallala Aquifer*

Source: Center for Geospatial Technology, *Impact of Alternative Water Policies on Rural Economies and Aquifer Hydrology.*

income earned in agricultural production is derived from crop production, and 81% of all crop revenue is derived from irrigated production.[26] Agricultural sales account for 27% of regional economic activity. When Lubbock County, the regional trade center with a population of about 220,000, is excluded, agricultural sales represent 79% of all regional sales.[27] The annual value of irrigated crop production for the four major field crops—cotton, wheat, corn, and grain sorghum—continues to have a significant economic impact on the THP economy. Annual gross receipts for these four crops totaled just under $2 billion in 2002,[28] and 84% of these receipts was generated from irrigated production. Cotton is the most important crop in the region, and that grown here accounts for 55% of all Texas cotton production.[29]

Other significant economic sectors in the THP are manufacturing, oil and gas, and trades and services.[30] Manufacturing activity tends to be concentrated in agricultural processing, oilfield equipment, and electronics. The trades and services sector includes wholesale and retail trade, finance, insurance, legal, business, advertising, medical, research, entertainment, repair services, and higher education. The agricultural industry has great impacts on the region's wholesale and retail trade, services, and financial sector.

CONTEMPORARY PLANNING IN THE THP

Reconciling the rule of capture with management institutions designed to conserve groundwater supplies has proven to be a very difficult task in the THP. Opponents of the rule of capture claim that technological development in hydrologic modeling in combination with an increasing water demand for limited groundwater resources have made the rule obsolete, whereas supporters of the rule claim that it protects the property rights of the landowners while allowing the Texas legislature to manage groundwater resources by creating local underground conservation districts. This section provides a brief overview of the rule of capture and the historical development of underground water conservation districts.

Property Rights

The Texas rule of capture doctrine facilitated the rapid expansion of irrigated agriculture. The doctrine places no significant restrictions on the volume of groundwater a landowner can withdraw from a well located on private property. By Texas law, the landowner has a right of capture derived from the English common-law maxim of property: *Cujus est solum, ejus est usque ad coelum et ad infernos* (To whomever the soil belongs, he owns also to the sky and to the depths).[31] This encourages the landowner to view groundwater as an exclusive private good. However, the rule of capture is more appropriately viewed as establishing a common property resource, because all landowners above the aquifer may pump from the aquifer, and aquifer withdrawals are not quantitatively partitioned among landowners in any way. Only individuals not owning land above the aquifer are legally prevented from withdrawing water.

Because of the common property nature of the Southern Ogallala Aquifer when governed by the rule of capture, a landowner cannot securely conserve water underneath his or her land for future use. Groundwater is reduced to ownership only when it has been pumped to the surface under existing legal doctrine (see Chapter 3). If the landowner conserves some water by leaving it in-ground, this conservation effort may be negated by the actions of other landowners who also have a right of use. Beattie described the problem as one of whether the aquifer functions as a "bathtub" or an "egg crate."[32] Alley and colleagues later modified the analogy to a comparison between a "bathtub" and an "egg carton."[33] In the egg carton construct, the irrigator's water supply is not seriously affected by a neighbor's pumping level, whereas in the bathtub scenario, the aquifer behaves like a lake in that each irrigator's pumping affects the water supply and pump lift of all other users.

If the aquifer were like an egg carton, with distinct compartments lying beneath each separately owned land parcel, then landowners would incur all benefits and costs of their actions with no third-party effects. Conservation effort thus would yield rewards for the conservator without the possibility of these rewards being undermined by other landowners. If, however, the aquifer behaved like a bathtub, then third-party effects would be unavoidably present. Alley et al. indicated that the Southern Ogallala Aquifer actually lies somewhere in between.[34] An individual user can affect a neighbor's pump lift, but the effect diminishes with distance, as is typical for aquifers worldwide. Therefore, the cumulative effect of many users over time does diminish water storage throughout the aquifer.

The Committee on Valuing Groundwater noted that individuals with groundwater access have little incentive to voluntarily conserve the water for future use when it is managed as a common property resource.[35] Groundwater not pumped by one individual is available to be used by competing individuals and thus is less likely to be conserved for future use. Therefore, competing users tend to ignore marginal user cost, which is the future opportunity cost of not having the resource available for use later because the water was withdrawn today. When marginal user cost is ignored, the true social cost of extraction in the current time period is undervalued, and the groundwater tends to be depleted too rapidly.[36]

Underground Water Conservation Districts

In 1949, recognizing that individual self-interest can result in the collective mismanagement of a common property resource such as an aquifer, the Texas legislature established a petition process for designating underground water conservation districts (UWCDs) with the objectives of conserving and managing groundwater. The High Plains Underground Water Conservation District #1, located in the Texas High Plains, became the first groundwater district in 1951.[37]

SB 1 identified these districts as preferred regional groundwater-planning groups and designated the Texas Water Development Board as the lead state agency in the regional water-planning effort. As a first step in the planning process, the TWDB divided Texas into 16 regional water-planning groups, as mentioned earlier. Region O, also referred to as the Llano Estacado Regional Water Planning

Region, accounts for nearly 96% of all Southern Ogallala Aquifer withdrawals in the THP. The 21 THP counties contained in Region O consist of Bailey, Briscoe, Castro, Cochran, Crosby, Dawson, Deaf Smith, Dickens, Floyd, Gaines, Garza, Hale, Hockley, Lamb, Lubbock, Lynn, Motley, Parmer, Swisher, Terry, and Yoakum. The High Plains Underground Water Conservation District #1 comprises 15 counties, 13 of which are part of Region O. The other 2 counties in District #1 are Randall and Armstrong, both of which only fractionally overlie the Southern Ogallala Aquifer.

The SB 1 planning process required each regional planning group to submit a management plan to the TWDB by September 1, 2000. These were to include both a drought and a water management plan for the region, consistent with the guidance principles for the state water plan. In 2000, the Region O planning group presented the TWDB with a report that included projections of expected future regional water use and supplies over a 50-year planning horizon (2000–2050). The report presented 16 water management strategies; of these, 13 were identified as short-term strategies for 2000–2030, and 3 as long-term strategies to be implemented in 2031–2050.

The short-term water management strategies consisted of enhancing groundwater supply sources for cities projected to need additional municipal water supplies, precipitation enhancement and weather modification, brush control, desalination of brackish groundwater, reuse of municipal effluent, municipal water conservation, irrigation water conservation, agricultural water conservation practices on farms, recovery of capillary water, cistern well construction, development of Lake Alan Henry reservoir, research and development of drought-resistant crops, and new technology. The long-term strategies were inter- and intrabasin basin water transfers, reuse of municipal effluent for potable water, and storm water capture, treatment, and use. The regional water plan was submitted to the TWDB for inclusion in the state water plan. The initial plan was updated in 2005, in accordance with the regional planning process, and submitted to the TWDB in September of that year.

SB 2 continues the direction of SB 1 by improving the planning, coordination, and management of Texas's water resources. This bill states that it is not the responsibility of groundwater districts to determine the appropriate location of the use of groundwater pumped from a well within a district. Specifically, districts cannot regulate water exporters more restrictively than in-district water users or adopt rules prohibiting the export of water, and they must remain objective in deciding on export permits. Districts must apply the same standard for wells used to export water out of the district as for wells used for in-district purposes. They can assess export surcharges within limits established in the Water Code.[38]

Article 2 of SB 2 authorizes groundwater conservation districts to assess production fees based on the volume of water withdrawn, but limits these fees to a maximum of $1/ac-ft for water used in agricultural production and $10/ac-ft for nonagricultural withdrawals. Production fees can be assessed instead of, or in addition to, any other taxes levied by the district.[39] Prior to the passage of SB 2, pumping quotas and well location were the only conservation policy tools available to groundwater districts. To date, no district in the THP has imposed either a quota restriction or a pumping fee to achieve a conservation goal. However, three major groundwater districts in the THP—the Panhandle Groundwater Conservation District, High Plains Underground Water Conservation District, and North Plains Groundwater Conservation District—have been debating the merits of establishing a conservation target that will conserve at least 50% of existing district-controlled groundwater reserves in aquifer storage 50 years into the future. The cost-effectiveness of using either a quota or pumping fee to achieve this potential conservation target has been investigated using the THP Water Policy Model discussed below.

THP WATER POLICY MODEL

The increasing scarcity of THP groundwater supplies has focused attention on the effectiveness of existing groundwater conservation management policies. Wantrup defines conservation as a postponing change in the intertemporal distribution of the physical rate of use.[40] Under this perspective, conservation means to shift resource use from more immediate time periods to future time periods. Thus, conservation is a relative concept that involves comparing two or more intertemporal use distributions.

In applied water policy analysis, the economic net benefit or cost-effectiveness of potential water conservation policies is normally estimated relative to the status quo baseline policy, which is established by using a water policy model to simulate water demands, groundwater reserves, and net economic returns over a specified planning horizon under existing technology, economic incentives, and water policy institutions. Once the baseline condition is established, water use and net economic returns are simulated over the specified planning horizon for a specific conservation policy. Potential conservation policies may include one or more policy tools: water withdrawal quotas, water fees (taxes), or a mandated increase in irrigation system efficiency. The annual stream of simulated net economic returns and water use associated with a given conservation policy are subsequently compared with the baseline condition to estimate the net present value economic cost of the conservation policy and the volume of groundwater conserved.

The THP Water Policy Model developed by Das and Willis is used to evaluate the conservation cost-effectiveness of the policy tools available to THP groundwater conservation districts.[41] The model links 19 county-level dynamic economic optimization models of irrigated production agriculture to a detailed spatially disaggregated hydrologic model of the Southern Ogallala Aquifer. This integrated economic and hydrologic model is used to study the conservation effect and agricultural policy cost of the potential water conservation policies over a 50-year planning horizon. Conservation effectiveness and policy cost are measured relative to a dynamic 50-year baseline established for each county under current economic incentives and hydrologic conditions.

This analysis uses Stovall's hydrologic model, which is calibrated for the entire Southern Ogallala Aquifer, spanning 29 counties in the Texas Panhandle and 5 counties in northeastern New Mexico. Stovall's model divides the land overlying the aquifer into a rectangular grid with 1-square-mile cells. The Southern Ogallala Aquifer grid consists of 246 rows and 184 columns, or 45,264 grid cells. Each grid cell contains parameter values for hydraulic conductivity, specific yield, recharge rate, initial saturated thickness, and the initial (current) volume of water withdrawn from each cell in the baseline calibration period.[42] The model for this analysis was constructed with the widely used MODFLOW groundwater simulation software program. Given user-provided parameter values for the aquifer's physical characteristics, MODFLOW employs a finite numerical difference equation procedure in combination with water budgets that account for recharge, withdrawals, and net lateral inflows to monitor saturated thickness and water table elevation through time.[43]

Detailed county-level economic models are constructed for only 19 of the 34 Texas and New Mexico counties that overlie the Southern Ogallala Aquifer. These 19 THP counties are shaded gray in Figure 9.3 and account for 97% of all Texas agricultural groundwater withdrawals from the Southern Ogallala Aquifer.[44] For 10 of the Texas counties that overlie the aquifer, economic models are not constructed because less than 10% of the county land area is above the aquifer or irrigated acreage is minimal. Additionally, because economic data for the 5 New Mexico counties are not readily available, no economic models are constructed for these counties. The annual water withdrawal level in each county for which an economic model was not constructed is maintained at the initial calibration level over the 50-year simulation period.

The economic model estimates the optimal agricultural groundwater extraction time path that maximizes the present value of agricultural net returns over a 50-year planning horizon. The Crop Production and Management Model is used to develop nonlinear crop production functions to describe crop yield response to applied water for given soil types, irrigation systems, and average weather conditions.[45] County-specific irrigated crop production functions are estimated for the five dominant irrigated crops grown in the 19-county study area (95 production equations in all). These crops are corn, cotton, grain sorghum, peanuts, and wheat and collectively account for 97% of agricultural crop water use within the study area. To provide a dryland alternative to irrigation, county-specific average dryland yields are determined for each of the crops, assuming average weather conditions and representative management techniques.

Additional county-specific data input into the dynamic economic model include initial saturated thickness, initial average pump lift, initial average well yield, initial average acres served per well, and initial number of irrigated and dryland acres by crop. The variable costs for dryland crop production and the additional costs for irrigation are taken from enterprise budgets for Texas Extension District 2.[46] Energy data include an energy use factor for electricity of 0.164 kWh/feet of lift/acre-inch, system operating pressure of 16.5 pounds per square inch, energy price of $0.0633 per kWh, and pump engine efficiency of 50%. Other costs include the per-acre cost of each irrigation system, irrigation system depreciation, annual per-acre irrigation system labor, and maintenance. A 3% real discount rate is used to convert 50 years of annual returns to present values. By linking the economic models to the hydrologic model, the integrated modeling approach is able to maintain the spatial variability in hydrologic response to agricultural groundwater stresses.[47]

In the policy analysis, it is assumed that each conservation policy tool is imposed with the objective of achieving the policy goal of maintaining at least 50% of initial groundwater reserves in aquifer storage at the end of a 50-year planning horizon in each THP county. A dynamic 50-year baseline for expected annual groundwater withdrawals is established for each county. The baseline status quo extraction time path is derived under the assumption that agricultural producers will optimally adjust annual groundwater withdrawals in response to increasing water scarcity over time, given current water policy regulations, private economic incentives, existing irrigation technology, and initial hydrologic conditions within each county. This efficient dynamic baseline condition subsequently is used as the frame of reference to analyze the conservation policy effectiveness of two conservation policies available to the two districts in the THP and a third tax policy not currently under consideration.

The first policy consists of imposing the legal maximum tax of $1/ac-ft on groundwater withdrawals over a 50-year planning horizon. The second is a quota policy on groundwater withdrawals designed to ensure that at least 50% of a county's initial groundwater reserves remains in storage at the end of the 50-year simulation period. Simulation results reveal that the $1/ac-ft tax rate fails to conserve 50% of initial groundwater reserves in 6 of the modeled 19 Texas counties. Moreover, the $1 tax has little effect in stimulating conservation behavior in any county. The third policy simulation, which is not currently being considered by either UWCD in the THP, estimates what conservation tax rate is required in each of these 6 counties to achieve this conservation goal.

Baseline Conditions

As reported in Table 9.1 for baseline conditions, the decline in irrigated acres, cumulative groundwater use, and ending groundwater reserves vary significantly by county. The ratio of the ending aquifer reserve level to the initial reserve level is used to estimate the degree of groundwater mining in each county. The smaller the computed ratio, the greater the degree of groundwater mining. In each county, the initial groundwater reserve level is calculated as the sum of all groundwater contained in each 1-square-mile MODFLOW aquifer cell with irrigation withdrawals in the calibrated first simulation year. Thus, the initial groundwater storage calculation for each county excludes the volume of water contained in any MODFLOW aquifer cell that currently does not supply water for irrigation use. Not all acreage above the aquifer with stored water is suitable for irrigated agriculture. This definition of initial reserves is used because the major focus of THP conservation policy to date has been toward curbing groundwater use by irrigated agriculture.

As shown in Table 9.1, at the end of the 50-year baseline simulation, regional groundwater storage is 61.5% of initial storage. Individual county reserves ranged from a high of 86.2% of initial storage in Bailey County to a low of 3.1% in Briscoe

Table 9.1 *Irrigated acres, cumulative groundwater withdrawals, and percentage of initial year 1 supplies in aquifer storage at end of 50-year simulation under baseline conditions*

County	Irrigated acres, year 1	Irrigated acres, year 50	Percentage of year 1 irrigated acres in irrigation in year 50	Cumulative groundwater withdrawals (1,000 ac-ft)	Ending percentage of initial reserve
Bailey	86,150	86,150	100.00	7,878	86.20
Briscoe[a]	31,236	1,790	5.73	1,839	3.10
Castro[a]	277,574	55,410	19.96	10,902	45.78
Cochran	86,812	18,961	21.84	3,458	80.36
Crosby	149,421	91,410	61.18	8,140	65.41
Dawson	46,919	27,654	58.94	2,748	78.60
Deaf Smith[a]	153,878	31,164	20.25	11,759	40.07
Floyd	197,906	30,205	15.26	10,775	81.02
Gaines[a]	271,142	76,195	28.10	12,210	46.40
Garza	12,366	7,162	57.92	564	84.29
Hale	338,347	88,884	26.27	12,017	75.61
Hockley	167,649	12,697	7.57	5,188	54.29
Lamb	243,772	78,885	32.36	12,266	73.02
Lubbock	217,120	17,018	7.84	6,624	58.46
Lynn	81,163	38,675	47.65	3,583	59.03
Parmer	233,942	7,581	3.24	4,843	85.08
Swisher[a]	39,808	8,361	5.98	3,094	37.11
Terry	177,070	42,257	23.86	5,325	52.32
Yoakum[a]	102,295	32,526	31.80	4,443	47.18
Region	3,014,570	752,984	24.98	127,658	61.46

[a] County with less than 50% of its initial groundwater reserves at the end of the 50-year simulation under baseline conditions.

County. In six counties, ending stored reserves are less than 50% of their beginning level.

Despite the significant county drawdown levels, the simulated baseline drawdown levels are significantly less than those previously predicted by Stovall, who used the TWDB projections for future agricultural groundwater withdrawals in his 50-year simulations.[48] A major limitation of the TWDB forecast demand estimates is that they were derived under the assumption that agricultural groundwater use would remain relatively constant over time, with agricultural producers maintaining all current irrigated acreage in production over the 50-year planning horizon. The TWDB future agricultural water demand estimates decrease only minimally over time in response to expected increases in irrigation efficiency. Thus, any decrease in the economic incentive to withdraw groundwater because of increasing cost is ignored in Stovall's analysis, and any simulated reduction in withdrawals through time is attributable solely to the inability of the aquifer to satisfy assumed agricultural use requirements because of the physical exhaustion of groundwater supplies in specific areas of the aquifer system.

In contrast, the change in the simulated aquifer withdrawal level through time in this study reflects both increasing economic cost of withdrawal and the physical exhaustion of available groundwater supplies in certain areas of the THP. From an economic perspective, an additional unit of groundwater will not be withdrawn when the marginal value of extracting the additional unit is less than the marginal cost of extracting it. Because the TWDB future use estimates ignore the increasing economic cost of pumping an acre-foot of groundwater as the water table declines, those estimates overstate future withdrawal levels and result in a simulated outcome that accelerates the groundwater depletion rate. Most areas of the THP will see significant reductions in the amount of land irrigated because of increasing pumping cost long before physical exhaustion occurs. In irrigated agriculture, economic exhaustion results when the cost of applying irrigation exceeds the additional revenue generated. Effective water policy management must accurately account for the changing economic incentive to withdraw groundwater as extraction cost and water scarcity change through time.

Under baseline conditions, a significant fraction of irrigated acreage is converted from irrigated to dryland production, in most counties, by the end of the 50-year simulation. The reduction is most noticeable in Briscoe, Hockley, Lubbock, Parmer, and Swisher Counties, where irrigated acreage decreases by more than 90% from its initial level, as reported in Table 9.1. Many counties experiencing a significant reduction in irrigated acreage had more than 50% of their initial groundwater reserves in storage at the end of the 50-year simulation. Again, this finding highlights the fact that economic profitability, and not just physical groundwater exhaustion, is an important factor in determining future irrigated acreage. Baseline results are derived under the assumption that producers will convert initial irrigated crop acreage to dryland cropping activities when either the groundwater supply is physically exhausted or economic exhaustion occurs as a result of increased pumping cost and dryland production becoming more profitable than irrigated production. Among the 19 counties, Bailey is the only county to keep all initial irrigated acreage in production over the 50-year simulation. Bailey County producers are able to maintain all irrigated acreage in production because they have a very profitable crop mix, plentiful groundwater supplies, and a low initial pump lift.

Over time, the increasing cost of lifting an acre-foot of groundwater because of water table declines will make it unprofitable to continue producing many crops under irrigation. Regionally, only 25% of all initial irrigated acreage remains in irrigated production 50 years into the future, given current technology, crop prices, and input prices. It is important to note that the baseline simulation did not consider the impact of potential subsidies to ethanol producers under the new emerging U.S. energy policy. Under a subsidized biofuels energy policy, the energy subsidies will increase the value of corn, a crop widely grown in the THP, and accelerate the mining of the aquifer beyond the simulated baseline value. Nor are potentially rising energy costs modeled, meaning that economic depletion of the Ogallala may occur earlier than what is simulated here.

From a policy perspective, a policy that further increases the cost of withdrawing increasingly scarce groundwater supplies, such as with a withdrawal fee or user tax, will further reduce the economic incentive to withdraw groundwater and enhance conservation efforts. The next section examines the effectiveness of implementing the maximum $1/ac-ft production fee (essentially a per-unit tax) on agricultural groundwater withdrawals to conserve groundwater supplies.

Groundwater Conservation under $1 Conservation Tax Policy

The intent of implementing a $1/ac-ft conservation tax on agricultural groundwater withdrawals is to decrease the agricultural incentive to withdraw groundwater. Despite the considerable tax revenues generated, the tax is not an effective policy tool for curbing agricultural groundwater use in the THP. As shown in Table 9.2, the tax reduces total regional groundwater withdrawals by only 0.33% relative to baseline use over the 50-year planning horizon. The reduction varies by county, ranging from a low of 0.01% in Hale County to a maximum of 1.99% in Castro County. Generally speaking, an inverse relationship exists between the value of water in agricultural production and the percentage of water conserved in each county under the tax policy. Counties having irrigated acreage in high-value crops such as peanuts and corn will pay the tax rather than significantly reduce their groundwater withdrawal level. Moreover, less than 50%

Table 9.2 *Reduction in groundwater withdrawals by policy scenario*

County	50% conservation tax rate ($/ac-ft)	Percent reduction relative to baseline		
		$1/ac-ft tax	50% conservation tax	Quota
Bailey	1.00	0.44	0.44	0.00
Briscoe[a]	95.03	0.40	44.44	44.44
Castro[a]	5.35	1.99	10.29	10.29
Cochran	1.00	0.10	0.10	0.00
Crosby	1.00	0.29	0.29	0.00
Dawson	1.00	0.78	0.78	0.00
Deaf Smith[a]	7.38	0.23	4.42	4.42
Floyd	1.00	0.05	0.05	0.00
Gaines[a]	67.08	0.14	1.42	1.42
Garza	1.00	0.29	0.29	0.00
Hale	1.00	0.01	0.01	0.00
Hockley	1.00	0.12	0.12	0.00
Lamb	1.00	0.05	0.05	0.00
Lubbock	1.00	0.49	0.49	0.00
Lynn	1.00	0.27	0.27	0.00
Parmer	1.00	0.83	0.83	0.00
Swisher[a]	48.40	0.14	3.70	3.70
Terry	1.00	0.08	0.08	0.00
Yoakum[a]	89.94	0.11	1.44	1.44
Region	—	0.33	2.77	2.61

[a] County with less than 50% of its initial groundwater reserves at the end of the 50-year simulation under baseline conditions.

of initial reserves remain in storage in Briscoe, Castro, Deaf Smith, Gaines, Swisher, and Yoakum Counties at the end of the 50-year simulation under baseline conditions. The overall conservation effect of the $1 tax policy is minimal when compared with total water use under the status quo baseline policy.

Groundwater Conservation under 50% Conservation Tax Policy

Because the $1 tax fails to significantly curb baseline agricultural groundwater withdrawals and conserve at least 50% of initial reserves in all counties, a more potent tax policy is also investigated. In each of the 6 counties mining more than 50% of their initial reserves under the legal maximum $1/ac-ft tax, the county tax rate was increased to the rate necessary to conserve 50% of initial reserves in storage. The tax rate is maintained at $1/ac-ft level in the other 13 counties. This tax policy is labeled the "50% conservation tax" in Table 9.2. Regional groundwater withdrawals are 2.77% less under the 50% conservation tax than in the baseline. This is an 8.4 times greater reduction than achieved under the $1 tax policy.

For the 6 counties requiring the 50% conservation tax, the tax rate ranged from a low of $5.35 in Castro County to a high of $95.03 in Briscoe County. The variation in the county-level tax rate is attributable to unique considerations existing within each county. For example, Briscoe County has a highly profitable irrigated crop mix in the initial simulation year, crops with above average per-acre water requirements, and limited groundwater supplies because of a below-average saturated thickness. Given this combination of baseline economic and hydrologic factors, nearly all available groundwater is exhausted by year 20, and only 3.1% of initial reserves remain in storage at the end of the 50-year planning horizon (Table 9.1). Thus, a very high conservation tax is required in Briscoe County to sufficiently decrease producer incentive to withdraw groundwater to maintain 50% of initial reserves in storage. For a given county, the appropriate conservation tax level is a complex function of the profitability of the crops grown, initial groundwater supplies, crop substitution options, and the number of acres under irrigation in the initial year of the simulation.

As anticipated, the ending reserve level in each of the 13 counties levying the $1/ac-ft tax under the 50% conservation tax policy is nearly identical to its ending reserve level when all 19 counties are levied the $1/ac-ft tax. The slight difference in the two ending storage levels results from increased lateral groundwater inflow into these 13 counties from the 6 adjacent counties that decrease withdrawals in response to the higher 50% conservation tax. The ending aquifer storage level as a percentage of initial stored reserves is 50% in all 6 counties assessing the higher conservation tax rate, which is significantly greater than their respective ending baseline percentages reported in Table 9.1.

Groundwater Conservation under 50% Quota Policy

The 50% quota policy represents a direct regulatory approach to achieving the conservation goal of conserving at least 50% of initial reserves in storage 50 years into the future. The 50% quota policy restricts groundwater withdrawals to ensure that at least 50% of initial storage is conserved in each county. As shown in Table 9.2, the quota policy does not affect water use in 13 of the 19 counties, because in the baseline scenario, the ending reserve level in these counties exceeded the targeted 50% conservation goal. The quota policy did reduce groundwater use below the baseline level in the other 6 counties, and the reduction in each county is identical to the reduction achieved under the 50% conservation tax policy. However, the regional reduction is only 2.61% under the 50% quota policy, versus the 2.77% reduction realized under the 50% conservation tax policy. The regional reduction is less under the quota policy because no tax is paid on groundwater use. Under the tax policy, the tax payment deters water use in the 13 counties not affected by the quota policy.

Both the 50% quota policy and the 50% conservation tax policy achieve roughly the same regional level of water conservation, but agricultural producers will prefer the quota policy because the agricultural cost is less. Under a quota policy, policy cost is limited to foregone producer profits, whereas under the 50% conservation tax policy, agricultural producer loss also includes the tax paid on each unit of groundwater withdrawn. The next two sections briefly examine the per-acre net present value cost and the per-acre-foot net present value cost of conservation for each of these policies.

Per-Acre Cost of Groundwater Conservation

Table 9.3 reports the per-acre net present value (NPV) cost for each conservation policy over the 50-year planning horizon. Per-acre policy cost is calculated as the NPV reduction for a given policy relative to baseline NPV. Under baseline conditions, per-acre NPV returns ranged from a low of $94.64 in Dawson County to a high of $2,275.23 in Yoakum County, and weighted per-acre average return for the 19-county study area is $980.79. The variation in county per-acre NPV is the consequence of a number of factors. Primary among them are the county's initial groundwater supply; initial water table elevation, which determines pump lift in the first simulation year; the initial crop mix; average rainfall pattern; and soil fertility. Differences in crop water requirements, prices, and production costs also influence profitability.

Regionally, the $1 tax policy imposes a $15.11 per-acre cost on irrigated agriculture, much less than either the $212.36 per-acre cost incurred under the 50% conservation tax policy or the $35.43 under the 50% quota policy. Even though per-acre policy cost is lowest under the $1 conservation tax, the policy is the least effective conservation policy. Regional groundwater use is decreased by only 0.33% relative to baseline use. The 50% conservation tax policy conserves more than 8 times as much groundwater as the $1 tax policy, but at a 14-fold increase in per-acre cost. The 50% conservation tax imposes a greater per-acre cost on counties with a higher baseline per-acre NPV return. A very high conservation tax is needed in Briscoe, Gaines, and Yoakum Counties to encourage producers to reduce groundwater use and convert irrigated acreage to dryland production. In these 3 counties, per-acre NPV cost of the 50% conservation tax policy exceeds $900 over the 50-year planning horizon.

Table 9.3 *Per-acre net present value cost by policy scenario*

County	50% conservation tax rate ($/ac-ft)	Baseline per-acre NPV ($/ac)	$1 tax ($/ac)	50% conservation tax ($/ac)	Quota policy ($/ac)
				Per-acre NPV cost (NPV decrease relative to baseline)	
Bailey	1.00	695.48	20.17	20.17	0.00
Briscoe[a]	95.03	1,665.52	14.20	931.89	474.97
Castro[a]	5.35	629.48	28.28	120.83	25.27
Cochran	1.00	1,067.68	11.24	11.24	0.00
Crosby	1.00	769.10	18.43	18.43	0.00
Dawson	1.00	94.64	4.24	4.24	0.00
Deaf Smith[a]	7.38	372.48	20.79	144.65	24.98
Floyd	1.00	1,117.75	21.27	21.27	0.00
Gaines[a]	67.08	1,915.08	19.76	1,144.47	129.87
Garza	1.00	816.65	7.39	7.39	0.00
Hale	1.00	1,566.78	25.07	25.07	0.00
Hockley	1.00	440.91	10.20	10.20	0.00
Lamb	1.00	944.75	21.23	21.23	0.00
Lubbock	1.00	643.35	15.58	15.58	0.00
Lynn	1.00	907.75	6.76	6.76	0.00
Parmer	1.00	280.69	8.30	8.30	0.00
Swisher[a]	48.40	963.07	8.38	367.68	9.53
Terry	1.00	1,657.90	11.00	11.00	0.00
Yoakum[a]	89.94	2,275.23	14.87	1,147.41	9.40
Region	—	980.79	15.11	212.36	35.43

[a] County with less than 50% of its initial groundwater reserves at the end of the 50-year simulation under baseline conditions.

Per-acre cost is considerably less under the 50% quota policy than under the 50% conservation tax policy. Policy cost is $0.00 in the 13 counties where the quota does not restrict water use below the baseline level. In the 6 counties where the quota restricts water use below baseline use, per-acre cost is considerably less for the quota policy than for the 50% conservation tax policy. Relative to the baseline regional per-acre NPV of $980.79, regional per-acre NPV is 1.54% less under the $1 tax policy ($15.11 per-acre cost), 3.60% less under the 50% quota policy ($35.43 per-acre cost), and 21.60% less under the 50% conservation tax policy ($212.36 per-acre cost).

Per-Acre-Foot Cost of Groundwater Conservation

To provide a more meaningful per-acre-foot policy cost comparison, the per-acre-foot NPV cost for the 50% conservation tax policy and the 50% quota policy are compared in Table 9.4. The cost comparison is limited to these two policies, because both conserve approximately the same volume of groundwater relative to the baseline. Calculated policy costs assume that irrigated agriculture is not

Table 9.4 *Net present value policy cost and conserved groundwater: 50% conservation tax policy versus quota policy*

	Total NPV cost ($)		Conserved water (ac-ft)		NPV cost ($/ac-ft)	
County	Tax	Quota	Tax	Quota	Tax	Quota
Bailey	4,399,893	0	34,645	0	127	NA
Briscoe[a]	92,630,236	49,146,759	1,202,995	1,202,995	77	41
Castro[a]	51,200,955	9,427,322	1,113,064	1,113,064	46	NA
Cochran	2,388,835	0	3,952	0	665	NA
Crosby	5,029,119	0	23,391	0	215	NA
Dawson	1,648,519	0	21,409	0	77	NA
Deaf Smith[a]	54,102,344	8,519,776	515,260	515,260	105	16
Floyd	7,122,231	0	5,089	0	1,400	NA
Gaines[a]	507,881,445	57,632,495	171,350	173,350	2,994	333
Garza	378,769	0	1,619	0	234	NA
Hale	8,522,293	0	612	0	13,925	NA
Hockley	3,774,586	0	6,128	0	616	NA
Lamb	7,442,908	0	5,651	0	1,317	NA
Lubbock	5,583,174	0	32,273	0	173	NA
Lynn	2,502,527	0	9,552	0	262	NA
Parmer	2,820,491	0	39,775	0	71	NA
Swisher[a]	109,337,206	2,766,895	112,719	112,719	970	24
Terry	3,930,434	0	4,217	0	932	NA
Yoakum[a]	238,468,679	1,952,509	65,985	65,985	3,614	31
Region	1,109,165,044	129,445,756	3,369,326	3,181,373	329	41

Note: NA = not appropriate; no cost is computed because no water is conserved relative to the baseline outcome under the quota policy.
[a] Counties where the conservation groundwater tax level needed to conserve 50% of initial county groundwater reserves is greater than the legal maximum tax of $1/ac-ft.

compensated for policy-induced taxes paid or reductions in groundwater use. (For the 50% conservation tax rates, see Table 9.3.)

On a regional per-acre-foot basis, the NPV cost of groundwater conservation is $329 for the 50% conservation tax policy versus $41 for the 50% quota policy. The regional NPV cost to irrigated agriculture for the 50% conservation tax policy is $1.1 billion, more than 8.5 times larger than the $129 million regional cost for the 50% quota policy. Despite the large difference in policy cost, groundwater conservation is minimally greater under the 50% conservation tax than with the 50% quota policy (3.37 million versus 3.18 million acre-feet).

CONCLUSIONS

The policy analysis reveals that the two conservation policies now being considered by the three UWCDs in the Texas High Plains will minimally decrease agricultural groundwater use below expected future use under existing water policy regulations, economic incentives, irrigation technology, and hydrologic conditions. Despite the considerable revenue generated by the $1/ac-ft withdrawal tax, the tax is an ineffective policy tool for curbing groundwater use. Regional groundwater use is reduced by only 0.33% under the $1 tax policy relative to the baseline use level. Among the study-area counties, water savings range from a low of 0.01% in Hale County to a high of 1.99% of baseline use in Castro County. Regionally, the $1 withdrawal tax lowers the present value of baseline net regional agricultural returns by 1.54%. The 50% quota policy reduces groundwater use by 2.61%, an 8.4 times greater reduction than achieved under the $1 tax policy, but at more than twice the agricultural cost. The 50% quota policy decreases the net present value of regional agricultural returns by 3.60%. These findings indicate that much more restrictive groundwater conservation policies must be considered if the policy intent of conservation is to significantly restrict agricultural groundwater use below expected future use levels.

In lieu of the 50% quota policy, a sufficiently high per-acre-foot conservation tax could be assessed on groundwater withdrawals to achieve the policy goal of conserving 50% of current groundwater reserves in each THP county 50 years into the future. Implementing the 50% conservation tax policy decreases the net present value of regional agricultural returns by 21.60%. This reduction is considerably larger than the 3.60% reduction incurred under the 50% quota policy, even though both policies conserve roughly the same volume of groundwater (2.77% decrease under the 50% conservation policy versus 2.61% decrease under the 50% quota policy).

Whether a tax or quota policy ultimately should be implemented in the THP depends on the management perspective. If the social objective is to maintain 50% of current groundwater reserves in situ in 50 years, and the policymaker is interested in minimizing the cost to irrigated agriculture, then the quota policy is the preferred choice. This presumes that administration and enforcement costs are the same under both policies. Some may argue that a quota is not a practical policy tool, because monitoring groundwater withdrawals throughout the region would present the state with an enormous monitoring and enforcement task. Yet a tax policy would require the same level of monitoring if the tax is collected. If administration and enforcement costs are high, a management agency may prefer a tax policy because it generates revenues that can be used to administer the conservation program. Moreover, the conservation management agency might wish to use any tax policy revenues generated in excess of administration costs to retire groundwater rights should a Texas legal modification establish true property rights in groundwater.[49]

Perhaps water policymakers in the THP should address the fundamental question of why groundwater conservation in the THP is in the social interest. Political decisionmakers in combination with the managers of the groundwater districts consistently argue that they are preserving scarce groundwater resources for future generations. However, these future uses remain unidentified. If the intended future uses pertain to nonagricultural (residential or commercial) demand within the THP, plenty of water may remain for these uses given available forecasts of population and consequent nonagricultural water demand, albeit at increasing cost over time.

If the intended future groundwater use is for irrigated production agriculture, then forcing irrigated agriculture to reduce its expected rate of groundwater withdrawal likely will provide no economic benefit to typical agricultural producers under current technology who are optimally managing their groundwater supply over the 50-year planning horizon. Given the lack of crop substitution options, postponing the rate at which agriculture extracts groundwater effectively depreciates the net present value of groundwater reserves that could be used in agricultural production if the conserved water is subsequently applied to the same crop mix. Additional advances in irrigation application efficiency would offset some of this conservation cost to irrigated agriculture, but given that well-designed LEPA center pivot systems have an application efficiency of 95% and drip irrigation systems have an application efficiency of 99%, the offsetting benefit of technological improvement is likely to be small. Although it remains unknown whether advances in genetic engineering will provide a means to increase crop yield response per unit of consumptively used irrigation water, such advances would further reduce conservation cost to irrigated agriculture and potentially provide a net economic return. Given that the model conducted here treats agriculture as being homogeneous in terms of crop selection and options within counties, it is likely some high-value producers would benefit from the conservation policies considered, at the expense of less profitable producers. However, the analytic model used in this policy analysis does not have sufficient spatial detail to address these potential distributional consequences.

If policymakers envision allowing agricultural landowners to sell their groundwater access to higher-valued residential and commercial uses in Texas's urban centers outside the THP, then decreasing agricultural groundwater withdrawal levels potentially could improve the economic welfare of irrigated producers, as well as increase social welfare. However, conventional groundwater districts in the THP do not embrace the concept of water markets allowing transfers from low-value agricultural uses to higher-valued residential and commercial uses in major population centers outside the region, such as Dallas or San Antonio. Policymakers must confront the tough political question of who is intended to benefit from groundwater conservation policies so that the value of conservation can be determined. To date, focus has been on estimating the agricultural cost of groundwater conservation in the THP. In order to determine the optimal conservation level through time, policymakers need estimates for both the cost and future benefit of conservation.

Clearly, water policy managers in the THP must confront and accept the hard reality that irrigated agriculture cannot be sustained at its current scale of operation. From a practical perspective, the Southern Ogallala Aquifer can be viewed as an exhaustible resource that has been mined to 50% of its original reserve level. If the policy purpose of groundwater conservation is to transform the regional economy from one based heavily on agricultural mining of groundwater to one that is less water-dependent, then the policy debate must be broadened to identify an alternative set of regionally preferred economic activities that are less water-intensive. Once the appropriate alternative economic future is identified, water policy research should then focus on identifying and implementing water conservation policies that will effectively transition the economy from heavy irrigation dependence to a regional economy consistent with a much smaller water endowment.

NOTES

1. Texas Water Development Board (TWDB), *Water for Texas 2007*, Doc. No. GP-8-1 (Austin, TX, 2007), available at http://www.twdb.state.tx.us/wrpi/swp/swp.asp (accessed June 15, 2010).

2. M.V. Guru and J.E. Horne, *The Ogallala Aquifer* (Poteau, OK: Kerr Center for Sustainable Agriculture, 2000), available at http://www.kerrcenter.com/publications/ogallala _aquifer.pdf (accessed June 15, 2010).

3. Ogallala Commons, October 10th Conference on Water in Ogallala, NE, press release, no date, http://ogallalacommons.org/docs/ogallala_pr1.doc (accessed July 13, 2004; site now discontinued).

4. 75th Texas Legislature, Regular Session, Texas Senate Bill 1, 1997, http://www.legis.state.tx.us/BillLookup/History.aspx?LegSess=75R&Bill=SB1 (accessed June 15, 2010).

5. 77th Texas Legislature, Regular Session, Texas Senate Bill 2, 2001, http://www.legis.state.tx.us/BillLookup/History.aspx?LegSess=77R&Bill=SB2 (accessed June 15, 2010).

6. Texas Joint Committee on Water Resources (TJCWR), *Interim Report to the 78th Legislature*, November 2002, http://www.senate.state.tx.us/75r/Senate/commit/c895/Down loads/water.pdf (accessed June 15, 2010).

7. TWDB, *Water for Texas 2007*.

8. House Research Organization (HRO), *Managing Groundwater for Texas's Future Growth*, Texas House of Representatives Focus Report, March 23, 2000, http://www.hro.house. state.tx.us/focus/ground.pdf (accessed June 15, 2010).

9. B. Das and D.B. Willis, Towards a Comprehensive Water Policy Model for the Texas High Plains, *Journal of Agricultural and Applied Economics* 36, no. 2 (2004): 510.

10. High Plains Underground Water Conservation District (HPUWCD), n.d., The Ogallala Aquifer, http://www.hpwd.com/Ogallala.asp (accessed June 15, 2010).

11. J.N. Stovall, Groundwater Modeling for the Southern High Plains (PhD diss., Texas Tech University, 2001).

12. HPUWCD, Ogallala Aquifer.

13. Stovall, Groundwater Modeling.

14. R. Nativ, *Hydrogeology and Hydrochemistry of the Ogallala Aquifer, Southern High Plains, Texas Panhandle and Eastern New Mexico*, Report of Investigations No. 177 (Austin, TX: Bureau of Economic Geology, 1988).

15. D.E. Green, *Land of the Underground Rain: Irrigation on the Texas High Plains, 1910–1970* (Austin, TX: University of Texas Press, 1973).

16. Ibid.

17. Ibid.

18. Ibid.

19. Ibid.

20. Ibid.

21. TWDB *Water for Texas 2007*.

22. J. Lewis, The Ogallala Aquifer: An Underground Sea, *EPA Journal* 16 (1990): 42–44.

23. A.R. Dutton, R.C. Reddy, and R.E. Mace, Saturated Thickness in the Ogallala Aquifer in the Panhandle Water Planning Area: Simulation of 2000 through 2050 Withdrawal Projections (Austin, TX: Bureau of Economic Geology, University of Texas, 2000).

24. U.S. Census Bureau, *2002 Economic Census*, 2002, http://www.census.gov/econ/ census02/ (accessed June 15, 2010).

25. U.S. Department of Commerce Bureau of Economic Analysis (BEA), Personal Income by Major Source and Earnings by Industry, Lubbock-Levelland, TX (EA), Table CA05N, 2004, http://www.bea.gov (accessed June 15, 2010).

26. Texas Agricultural Statistics Service (TASS), *2002 Texas Agricultural Statistics* (Austin, TX: TASS, 2003).

27. BEA, Lubbock-Levelland, TX.

28. TASS, *2002 Texas Agricultural Statistics*.

29. Ibid.

30. HDR Engineering, Inc., *Llano Estacado Regional Water Planning Area Regional Water Plan* (Lubbock, TX: HDR Engineering, 2001).

31. R.A. Kaiser and F.F. Skillern, Deep Trouble: Options for Managing the Hidden Threats of Aquifer Depletion in Texas, *Texas Tech Law Review* 32, no. 2 (2001): 263.

32. B.R. Beattie, Irrigated Agriculture and the Great Plain: Problems and Policy Alternatives, *Western Journal of Agricultural Economics* 6, no. 2 (1981): 292.

33. W.M. Alley, T.E. Reilly, and O.L. Franke, *Sustainability of Ground-water Resources*, USGS Circular 1186 (Denver, CO: U.S. Geological Survey, 1999), 48.

34. Ibid.

35. National Research Council, Committee on Valuing Ground Water, *Valuing Ground Water: Economic Concepts and Approaches* (Washington, DC: National Academy Press, 1997).

36. R.C. Griffin, *Water Resource Economics: The Analysis of Scarcity, Policies, and Projects* (Cambridge, MA: MIT Press, 2006).

37. HRO, *Managing Groundwater*; O.W. Templer, Adjusting to Groundwater Depletion: The Case of Texas and Lessons for the Future of the Southwest, in *Water and the Future of the Southwest*, ed. Zachary A. Smith, 247–68 (Albuquerque, NM: University of New Mexico Press, 1989).

38. TJCWR, *Interim Report to the 78th Legislature*.

39. Ibid.; Texas Senate Bill 2.

40. S.V. Ciriacy-Wantrup, *Resource Conservation: Economics and Policies*, 3rd ed. (Berkeley, CA: University of California Division of Agricultural Sciences, Agricultural Experiment Station, 1968).

41. Das and Willis, Comprehensive Water Policy Model.

42. Stovall, Groundwater Modeling.

43. M.G. McDonald and A.W. Harbaugh, *A Modular Three-Dimensional Finite-Difference Ground-Water Flow Model*, USGS Water Resources Investigations Report 83-875, Techniques of Water-Resources Investigations, Book 6, 1988, http://pubs.usgs.gov/twri/twri6a1/html/pdf.html (accessed June 15, 2010).

44. Das and Willis, Comprehensive Water Policy Model.

45. T. Gerik, W. Harmon, J. Williams, L. Francis, J. Greiner, M. Magre, A. Meinardus, and E. Steglich, *User's Guide: CroPMan (Crop Production and Management Model)* (Temple, TX: Blackland Research and Extension Center, 2003).

46. Texas Agricultural Extension Service, *Texas Crop and Livestock Budgets: District 2*, 2003, http://agecoext.tamu.edu/resources/crop-livestock-budgets/ (accessed May 15, 2003).

47. A complete discussion of the THP Water Policy Model is found in B. Das, Towards a Comprehensive Regional Water Policy Model for the Texas High Plains (PhD diss., Texas Tech University, 2004).

48. Stovall, Groundwater Modeling.

49. See Chapter 4 for a discussion of the partitioning of groundwater required for a true private property rights system. Because recharge (flow) to the Ogallala is small, it may not be important to engage this part of the system, leaving partitioning of the stock (already in storage) as the more significant element.

CHAPTER 10

Advanced Technologies for Tapping Unconventional Texas Waters

David Jassby, Andrew J. Leidner, Yao Xiao,
Andreas Gondikas, and Mark R. Wiesner

*F*urther population increases are anticipated for Texas in the coming decades. In many of the same places experiencing rapid population growth, the traditional and most accessible freshwater supply sources are fully tapped. Confounding this issue, average yields from traditional water sources may decrease as a result of the accumulation of sediment in reservoirs and the depletion of aquifers. The forces of demand growth and possible supply attrition compel state and regional water managers to investigate innovative, previously unconsidered sources of fresh water, as well as alternative methods of controlling water demand, such as usage restrictions and higher prices. This chapter discusses several advanced technologies and projects that Texas water groups have employed, and may employ in the future, to acquire additional sources of fresh water for a thirsty and growing population.

For this discussion, the various advanced technologies are organized into three groups, based on the type of source water used by the technology: (1) desalination, (2) water reuse, and (3) water harvesting and fabrication. Each of these categories contains several technologies, ranging from an ancient approach, rainwater collection, to cutting-edge engineering innovations such as forward osmosis.

Desalination technologies produce fresh water from salt water, making available for exploitation an arguably limitless resource: the ocean. Advances in wastewater treatment and reuse have the potential to reduce both the contaminant load into the environment and the reliance on conventional water sources. Especially in rural and small communities, rainwater and humidity harvesting, which require relatively smaller amounts of capital to implement, may contribute more readily to local water supplies. The fuel cell, an emerging technology, holds the promise of supplying both clean energy and water, which is made all the more promising because of the interdependence of many water and energy production processes.

A water body that is unlimited in physical availability, however, does not imply a water source that is unlimited in economic availability. Each technology discussed in this chapter has a price tag in terms of capital costs, energy costs, operational costs, risk, and in some cases, waste disposal. Some of these costs, such as capital and energy costs, are readily available in published literature. Others, such as operational costs and, in particular, waste disposal costs, are not. This chapter discusses the energy requirements of each technology and provides several examples of capital costs of various installations around the world. These costs are unique to the region where they are located, however, so they may not be fully applicable to Texas. Also, the costs associated with the production of water are only part of the total cost that consumers pay. Additional charges associated with supplying potable water, such as disinfection and pumping, are not discussed here.

DESALINATION

The process of desalination, where dissolved solutes are separated from saline water to produce fresh water, is commonly used in regions around the world where seawater or brackish waters are readily available and freshwater sources are scarce. In addition to treating seawater or brackish water, many desalination technologies can be used in a water reuse context, where concentrations of dissolved salts must be reduced. Desalination processes can be organized into two groups: those based on thermal separation, or distillation, and those based on membrane separation.

Distillation Processes

In distillation, the feed stream of water is heated in a fashion that favors vaporization. Water vapor is then condensed and collected, and the resulting distillate is virtually salt-free. Desalination by distillation requires relatively little pretreatment of feedwater compared with membrane processes. However, the trade-off is that energy consumption per unit of water produced is greater for distillation than with most membrane processes. In addition to mineral salts, many organic solutes may also be removed by distillation, but the possibility exists for some volatile fractions of organic matter to remain in the distillate. Technological variations within distillation processes differ largely by the manner in which the feed stream heating or vapor condensation is performed.

Multistage Flash Evaporation. In 1999, multistage flash (MSF) evaporation processes were responsible for approximately 60% of the world's total desalination capacity.[1] The several types of MSF systems include circulation (MSF), once-through (MSF-OT), and mixing (MSF-M). The first stage of the MSF configuration involves saline feedwater flowing inside tube-shaped heat exchangers, with steam flowing on the outside of the tubes. The steam raises the temperature of the saline feedwater, which then moves into a series of chambers, or stages, with progressively lower pressures. The pressure in each stage is lower than the saturated vapor pressure of the saline water at that temperature. The process of evaporation in these chambers is known as flashing.

The flashing process results in a pressure gradient between the chambers, allowing the heated saline water to flow in a particular direction without the aid of pumps. The heat released by the cooling vapor is recycled by running the vapor through tubes that are surrounded by recycled saline feedwater that was not vaporized in the previous flashing stage, thereby conserving energy. After the last stage, any saline water that was not distilled in the previous stages is mixed with new saline feedwater, and the process is repeated. Intake saline feedwater is introduced to the system around the condensing tubes in the last flashing chamber, absorbing any latent heat from the vapor. This saline feedwater stream is then split into two streams, the first being discarded back into the ocean as a concentrate discharge stream, and the second being added to the recycled saline feedwater for another pass through the distillation stages.[2]

The MSF-OT configuration is essentially identical to the recycling configuration, except that all the saline water left from the flashing process is released back into the ocean. The MSF-M configuration differs from the recycling configuration only in that the entire intake stream is mixed with the recycled brine in a special chamber called the mixer, thus eliminating the cooling stream found in the MSF configuration (the stream that is discharged).[3] Multistage flash desalination plants are often coupled with power plants. By doing so, steam generated from the cooling water used in the power plant can be used to drive the desalination process.

Intake saline water typically is screened to remove large particles and then filtered. Feedwater to the MSF is deaerated, and antiscaling and antibubbling chemicals are added. Because the MSF-OT configuration discharges all the saline water left from the desalination process, chemical consumption in this configuration tends to be high. In contrast, the MSF and MSF-M configurations conserve chemicals by reusing the leftover brine. Because the MSF-OT does not recycle feedwater, water to be desalinated is typically of a lower salinity than water that is to be desalinated using the other configurations. This in turn makes the desalination process easier, as water with higher salinity has a higher boiling point. However, this advantage is offset by the fact that the MSF-OT system loses heat when the heated saline water is discharged. The inherent disadvantages of the MSF-OT system makes it uneconomical to operate on an industrial level.[4] The elimination of the cooling stream in the MSF-M process makes this configuration more energy-efficient than the MSF, as all of the heat from the final flashing stage is absorbed by the incoming water. MSF is the most energy-intensive desalination process. Typical energy inputs required are on the order of 51.7 kilowatt-hours per cubic meter (kWh/m^3).[5]

Multiple Effect Evaporation. In multiple effect evaporation (MEE), intake saline water is sprayed over plate- or tubelike heating elements filled with steam, in a series of chambers known as effects. Heating elements in the first effect are heated by externally supplied steam. In the first effect, some of the water that comes into contact with the heating elements evaporates and is collected. The remaining saline water is sprayed into the next effect, where the steam generated in the previous effect is now used to heat the heating elements. This process is repeated multiple times in a series, each step utilizing the latent heat of the steam generated in the previous effect. Steam generated in the last effect is used to heat the intake saline water.

In comparison with MSF, the relatively lower temperatures for operation of the MEE process (158°F [70°C]) reduce the potential for scaling in the system. Also, the MEE technology allows for shorter start-up times and offers greater flexibility in matching water production to quantity demanded.[6] Typical energy input values are in the range of 45 kWh/m^3 for MEE.[7]

Mechanical Vapor Compression. In mechanical vapor compression (MVC) desalination, intake salt water is sprayed on electric plate heat exchangers. As the water temperature rises, vapor forms, which is collected and subsequently compressed by a mechanical compressor and thus becomes superheated. This superheated vapor flows through condenser tubes surrounded by the intake salt water, releasing the heat to the intake water and condensing into the desalinated product. The leftover concentrated saline water is released back into the environment.

Typical MVC desalination units are operated entirely by electrical power and are best suited for smaller-scale operations. These systems are particularly suitable for remote locations that do not have access to the large quantities of fossil fuels needed to generate the steam used in other distillation technologies. The MVC systems can be constructed with multiple vaporization chambers, also known as effects. This increases the capacity of the systems, allowing for larger volumes to be treated, as needed.[8] MVC requires relatively little energy to operate. Energy input ranges between 10 and 11 kWh/m^3.[9]

Geothermal. Another small-scale desalination system, useful in areas with an abundant supply of geothermal brackish springwater, was developed by Caldor-Marseilles in 1994.[10] In this system, geothermal brackish water is introduced into tubes at the top of an evaporator unit and is allowed to flow through them to a collection basin at the bottom of the unit. While this water flows through the tubes in the evaporation unit, it passes its heat to its surroundings and cools down. Air is pumped into the evaporation unit, flowing upward around the tubes containing the hot brackish water. Some of the cooled brackish water collected at the bottom of the evaporation unit is then pumped through tubes running up the condenser unit, absorbing heat from the surroundings. This water is sprayed onto the heat-exchanger tubes filled with hot brackish water in the evaporation unit, causing some evaporation to take place. The water vapor travels up through the evaporation unit, aided by air. This vapor then flows to the condenser unit, where water condenses on the outside of the tubes carrying the cooled brackish water, transferring the vapor heat to the cooled water in the tube. The condensed salt-free water is collected at the bottom of the condenser unit.

These units can be designed to supply small amounts of water, thus requiring very little energy, as a unit has only two water pumps and an air blower that need to be operated. Another advantage of this process is its use of temperatures that are low relative to other geothermal temperatures (158–194°F [70–90°C]). This range of operating temperatures allows for inexpensive polymers to be used in project construction. Although the geothermal resources in Texas are modest, areas in central and south Texas may be compatible with this technology.[11]

Solar Humidification.   In contrast with geothermal resources, the potential for solar energy coupled with desalination in Texas is substantial. The use of solar energy for the purpose of desalination can be categorized as direct or indirect. Indirect solar energy refers to solar radiation that is harnessed and converted to electricity, which is then used to power desalination processes such as the MSF and MEE. As solar panels become more efficient and affordable, it may become cost-effective to use solar-derived electricity for desalination purposes. As this is not currently the case, this section focuses on desalination applications that use solar energy directly. The simplest form of direct solar desalination, known as a solar still, consists of a basin full of saline water, with a slanted roof made of a transparent material covering the basin. Sunlight heats the water, causing it to evaporate. The water vapor then collects and condenses on the slanted roof, sliding into collector units located alongside the basin. The efficiency of the solar still has been enhanced primarily by improving the absorbance of the solar energy by the water. This has been accomplished by adding a dark dye to the saline water, painting the basin a dark color, or adding charcoal to the basin.[12] The solar still is very easy to construct and maintain, but it produces relatively small amounts of fresh water, approximately 3 to 4 liters per square meter (L/m^2) per day. This process can be improved with heat recycling,[13] where solar energy is used to heat oil, which in turn is used to heat brine in the first stage of desalination. As water vapor forms and condenses, the latent heat of the vapor is used to heat more brine, which causes more vapor to be created and collected. The process is repeated multiple times. This system is capable of producing up to $25 L/m^2$ per day. The quantity is highly dependent on solar intensity, however, and thus this technology is applicable only in regions with abundant sunshine.

A novel way of utilizing solar energy for desalination was described by Chafik.[14] In this system, air is passed through a series of solar-heating and humidification steps. In each step, the air is first heated through a solar collector, and then passed into a humidifier where saline water is sprayed into it, saturating the air with moisture. This in turn causes the air temperature to drop. The process is repeated in the next solar collector–humidifier stage, and the amount of moisture in the air increases further. Thus, very high moisture concentrations can be reached, up to 148 grams of water per kilogram of air after the fifteenth heating-and-humidification stage. The heated air from the last stage is then allowed to cool in a condenser, releasing the entrained moisture, which is collected as fresh water. The sequential heating of the humidified air allows for higher levels of moisture to be carried by this air, reducing the size of pipes and blowers needed to transport the air through the process. Typical energy inputs into a solar desalination unit are in the range of $0.3 \, kWh/m^3$, making this the most energy-efficient process, although it is highly dependent on the availability of solar energy.[15]

Distillation Technologies in Texas. Distillation technologies have not been broadly implemented in Texas for a number of reasons. The energy requirements for MSF and MEE are not trivial, as discussed in the relevant sections above. Texas has two MSF installations, one quite large (operated by Union Carbide in Texas City).[16] The development of geothermal desalination projects may be constrained in Texas because of the limited availability of geothermal brackish groundwater. Recently, MVC and solar distillation technologies have received attention and funding in the state.

Water scarcity within the city of Laredo, located in the Texas Middle Rio Grande Valley, has prompted city officials to consider construction of a brackish groundwater desalination facility to augment existing water supplies, which are primarily being drawn from the Rio Grande. A pilot-scale brackish groundwater desalination facility for Laredo is in early stages. This facility will test a relatively new vapor compression technology, developed in Texas, called advanced vapor-compression evaporation (AdVE).[17]

Another distillation technology that has received attention is the solar still. Foster and colleagues have reported on the progress of making solar distillation available to more than 200 families along U.S.–Mexico border. These facilities are primarily small-scale and used to augment the water supply for impoverished areas, called colonias, along the border.[18] Also along the Texas–Mexico border, the El Paso Solar Energy Association is raising awareness of solar energy applications, including solar distillation.[19] Finally, a project of the Rio Grande Basin Initiative is exploring solar distillation technologies as a potential source of irrigation water.[20]

Membrane Processes

Membrane desalination units are fast becoming the most prevalent desalination technology in the world. As of 2006, 38 public water supply entities in Texas were using membrane desalination processes.[21] The main advantage of membrane desalination systems is the relatively low amount of energy needed to operate them; their main disadvantage is the extensive pretreatment required for the feedwater entering the primary membrane treatment units.

Reverse Osmosis and Nanofiltration. Reverse osmosis (RO) and nanofiltration (NF) are pressure-driven membrane processes capable of removing significant quantities of mineral salts and, to a large extent, organic solutes. NF differs from RO primarily in that smaller-molecular-weight materials and monovalent salts are not removed. Essentially, the permeable pores of NF membranes are larger than those of RO membranes. In these processes, feedwater is introduced at a pressure greater than that of the osmotic pressure of the feed stream (up to 1,200 pounds per square inch for seawater desalination), inducing flow across a membrane that selectively rejects the dissolved salts. RO desalination is capable of removing up to 99% of solutes in the water, including monovalent ions such as chloride (Cl^-) and sodium (Na^+).

RO and NF membranes often are fabricated as thin film composite membranes, composed of a thicker supporting layer and a thin rejecting layer. The layer exposed to the feed stream, known as the active layer, is typically very thin and dense, and it is where the actual separation takes place. The chemistry of the active layer is vital in determining membrane performance, and many polymeric formulations have been commercialized. These membranes typically are assembled as spiral-wound or hollow-fiber elements.

In the spiral-wound configuration, multiple layers of two flat membrane sheets glued on both sides of a permeate collector material sheet, with each layer separated by a feed channel spacer, are wound around a permeate collection tube to form an elongated, tubelike membrane element. Pressurized feedwater is introduced at one end of the tube, and as the feed stream flows across the membranes, water diffuses through the membrane and is carried by the permeate collector material to the permeate collection tube. Typically, several membrane elements are connected in a series. As water diffuses from the feed stream, solute concentrations increase, and the feed stream continues to the next membrane element. Feed stream pressure drops as it travels along the membrane units.

The hollow-fiber membrane consists of a bundle of long, thin, tubelike membranes. The active membrane layer is on the outside surface of the tube. Feedwater typically flows from the outside of the membrane to the inside, where permeate flows along the length of the hollow fiber and is collected. A single hollow-fiber element in a bundle may have a diameter similar to that of a spaghetti noodle.

RO and NF membranes separate solutes from water through a number of mechanisms. Molecules may first partition into the membrane and then diffuse through the material to pass to the other side. Separation plays on the differences in solubility and diffusivity of molecules in the membrane material. NF membranes may have spaces within the membrane matrix that allow for convective flow of water. Therefore, straining of molecules larger than the effective size of the NF membrane pores is a second possible phenomenon that determines membrane rejection. These membranes typically are negatively charged because of the presence of functional groups that coordinate water, enabling water molecules to bind more easily to the membrane and allowing a greater flux of water across the membrane.

Several phenomena impair the performance of RO and NF membranes. The first is concentration polarization, where solutes accumulate near the membrane surface. As water diffuses across the membrane and solutes are rejected, local solute concentrations near the membrane surface increase. This leads to a steeper concentration gradient across the membrane, lowering the flux of water across the membrane. Also, as solute concentration increases, solute rejection decreases, and thus separation is impaired. To prevent concentration polarization, adequate velocity must be maintained in the feedwater, so as to promote mixing of the rejected solutes and prevent high local concentrations near the membrane.

Another problem frequently encountered in membrane desalting is membrane fouling. This occurs when materials accumulate on the membrane surface as well as within the membrane pores, and this accumulation impedes flow across the membrane. Membrane fouling may result from scaling or biological growth. Scaling occurs when solute concentrations reach high enough levels to precipitate onto the membrane surface, clogging it. Several techniques are used to prevent scaling: antiscaling chemicals can be added to the feed brine, pH levels can be adjusted to minimize scaling, and feed recovery can be limited so as to keep solute concentrations at an acceptable level. Biological fouling occurs when microorganisms attach and grow on the membrane surface. Flow rates across the membrane should be kept high enough to prevent attachment and growth of microorganisms. Also, biocides can be added to the feedwater to control microbial growth. However, chlorine, the most commonly employed biocide, cannot be used in conjunction with polyamide membranes and should be used only at very low concentrations with cellulose acetate membranes.

A typical RO or NF system consists of a series of membrane elements, each membrane receiving water from the previous membrane. Permeate is collected from each membrane element and removed from the system. Concentrate is eventually discarded, often requiring special disposal because of high salt concentrations. Usually, several membrane elements are connected in parallel, with the number of elements in parallel decreasing as water moves from element to element. This is done so as to maintain adequate flow through the system. The number of elements in a series is limited by the increasing concentration of solutes in the concentrate and the pressure drop.

Typical energy inputs into a commercial-scale RO desalination system are approximately $5.7 \, \text{kWh/m}^3$.[22] The largest RO plant in the world, a seawater desalination facility, is located in Ashkelon, Israel, and is designed to produce $330,000 \, \text{m}^3$ per year. A manager at the Dow Chemical Company, which provided the membranes to the Ashkelon facility, stated that the facility operates at a cost of $52 \, \text{cents/m}^3$, underscoring an industrywide trend of decreasing costs for the RO systems.[23]

Electrodialysis. Electrodialysis (ED) removes charged molecular species from water as they migrate in response to an applied electrical gradient. Positive ions (cations) are drawn to the negative electrode, and negative ions (anions) are drawn to the positive electrode. On their way to the electrodes, ions encounter two types of membranes. Cations traveling toward the negative electrode from the center of the vessel first encounter a cation exchange membrane, allowing them to pass through. Next the cations encounter an anion exchange membrane, which prevents them from reaching the negative electrode. Thus, the cations are trapped between the two membranes. The process is similar with the anions traveling toward the positive electrode, but in this case the first membrane encountered is an anion exchange membrane and the second is a cation exchange membrane. The result of this process is that water located in alternating membrane channels has a reduced ion concentration.

Ion exchange membranes gain their unique electrical properties through the presence of fixed charged functional groups embedded in the membrane polymeric matrix. Pores are present throughout the membranes. Cation exchange membranes have fixed negative charges, which assist cations in negotiating the pores while preventing anions from passing through. Anion exchange membranes have fixed positive charges, which facilitate the passage of anions and prevent the passage of cations. To reduce concentration polarization, the efficiency-reducing buildup of ions near membranes, the polarity of the electrodes is periodically alternated in a process called electrodialysis reversal (EDR). Feedwater for EDR units needs similar pretreatment to that required in the RO process.

Because it is ions that move across the membranes, rather than water as in RO or NF, there is no removal of pathogens or uncharged organic molecules. Typical energy input to an electrodialysis desalination unit is in the range of 7 to $10\,kWh/m^3$, depending on feedwater salinity.[24]

Forward Osmosis. In the process of forward osmosis (FO), differences in osmotic pressure are used to separate solutes from the feed stream. To induce the migration of water across the separation membrane, a draw solution is separated from the brine by a salt-selective membrane. The high concentration of the draw solution induces water to diffuse from the brine, while the membrane prevents the solutes from the two solutions from mixing. Very large differences in osmotic pressure can be developed by choosing appropriate solutes in the draw solution. The solute for the draw solution must be a material that can be easily removed, such as alcohol. Energy requirements of a FO plant are estimated to be in the range of 20 to $30\,kWh/m^3$.[25]

The draw solution is composed of brine plus an added set of chemicals. Once the draw solution is dilute enough, the added chemicals are removed, yielding potable water. Added chemicals include volatile compounds, such as sulfur dioxide or alcohols; chemicals that are easily precipitated, such as aluminum salts; or digestible compounds, such as sugars. A novel draw solution adds ammonium bicarbonate (NH_4HCO_3) to the saline water, and when dilution is finished, the draw water is heated ($140°F$ [$60°C$]), facilitating the breakup of the added chemical and causing carbon dioxide and ammonia (CO_2 and NH_3) gases to be released from the draw solution.[26]

Membrane Distillation. The development of alternative membrane processes that do not rely on pressure as a driving force have shown preliminary promise in the treatment and reuse of concentrated brine streams, significantly reducing or in some cases eliminating the need for concentrate disposal.[27] One such treatment alternative is the use of thermally driven membrane processes, also known as membrane distillation (MD). MD is a separation process that simultaneously combines mass and heat transfer through a hydrophobic microporous membrane.[28] In this process, water is evaporated across a membrane in response to a temperature differential. The driving force for mass transfer in this process is the vapor pressure difference between the two streams (i.e., the feed and permeate or distillate).

The hydrophobic nature of the membrane prevents the aqueous solution in the liquid phase from entering the pores, resulting in a liquid-vapor interface at each pore entrance. One of the more attractive attributes of thermally driven membrane processes is the relatively minimal impact of salt concentration in the feedwater on membrane performance.[29] Integrating thermally driven membrane systems into traditional membrane process configurations may result in significantly increased product water recoveries.[30]

Membrane Process Technologies in Texas. Desalination through membrane process technologies has been a growing component of Texas water supply planning and implementation in recent years. In the most recent state water plan, 8 of the 16 regional planning groups across the state include desalination as a recommended water management strategy, with particular interest given to RO desalination of brackish groundwater.[31]

Seawater desalination using RO has been proposed as a partial solution to industrial needs in a current Texas Water Development Board (TWDB) plan that examined the feasibility of three pilot projects along the coast. Located in Freeport (near Houston), Corpus Christi, and Brownsville, each of these facilities would have a capacity of at least 25 million gallons per day (MGD). The Brownsville site was deemed the most feasible, and a pilot study was conducted from February 2007 through July 2008. The study concluded that construction of a 25 MGD seawater desalination facility would cost approximately $152 million. The final report recommended a phased approach to developing the 25 MGD desalination facility, which would start with a demonstration facility that would produce 2.5 MGD at a facility cost of $67 million, with about half of those costs due to scaled-up components to ease future expansion of the facility. The additional capacity would be brought online as the area's demand for freshwater supplies increased over the life of the facility.[32]

Shown in Figure 10.1, 38 desalination facilities with a capacity greater than 25,000 gallons per day are operating in Texas. The majority of these facilities use RO, but some use EDR for desalination.[33] In 2005, the TWDB published a database that provides information about each desalination facility in Texas, including the year of construction, the quality of feedwater, and the supplier of RO membranes.[34] Since the desalination database was completed, the largest inland desalination facility in the world was constructed in El Paso, Texas. This RO facility, called the Kay Bailey Hutchinson Desalination Plant, is capable of producing 27.5 MGD of fresh water at full capacity.[35]

NRS Consulting Engineers, a water treatment and desalination engineering firm in Texas, provides an information service website about desalination activities in Texas.[36] This website is a portal to two significant Texas RO desalination resources. The first is the webpage of the Rio Grande Regional Seawater Desalination Project. This project brought together efforts from the Brownsville Public Utilities Board, the Port of Brownsville, the TWDB, and NRS to evaluate the technical feasibility of seawater RO along the Brownsville Ship Channel. This was the first seawater RO desalination pilot project in the state. The second

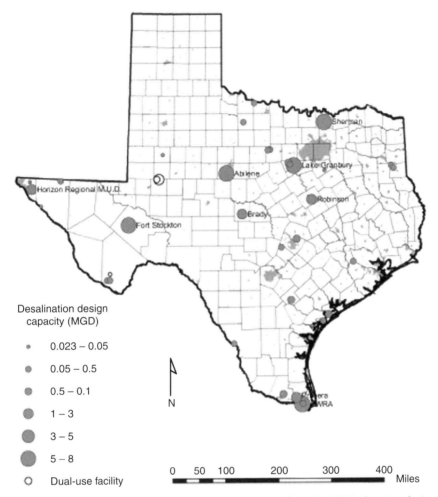

Figure 10.1 *Map of desalination facilities in Texas through 2005 showing design capacity*

Note: Facilities with a desalination capacity of at least 1.5 million gallons per day (MGD) are named; location of some facilities approximate.
Source: J.P. Nicot, S. Walden, L. Greenlee, and J. Els, *A Desalination Database for Texas* (Austin, TX: Texas Water Development Board, 2006).

resource at the NRS desalination portal is the website for the *Guidance Manual for Brackish Groundwater Desalination in Texas*. This manual, which was also funded by the TWDB, provides guidance to community water suppliers considering the use of brackish groundwater treatment technologies to augment their water supplies. The manual draws from the experiences of the North Cameron Regional Water Project, a recently completed 2.25 MGD brackish groundwater desalination facility in the Texas Lower Rio Grande Valley.[37]

Since the completion of the Rio Grande Regional Seawater Desalination Project, a second seawater RO desalination pilot project, the Laguna Madre Seawater Desalination Pilot Plant Study, has been constructed on South Padre Island. This project, a collaboration among the Laguna Madre Water District, TWDB, and NRS, is designed to collect source water quality data and determine the cost of a small-scale seawater RO desalination facility.[38] The Innovative Water Technologies (IWT) group of the TWDB is very involved with state water desalination activities. The IWT's desalination webpage is a gateway to many of the group's completed and ongoing desalination projects.[39]

Little information is available regarding the use of ED and EDR in Texas. A facility located southwest of Fort Worth and under the management of the Brazos River Authority, the Granbury Surface Water and Treatment System, had implemented EDR at one time but has since converted to an RO facility.[40] In addition to the Granbury facility, four other EDR facilities were listed in the Texas desalination database.[41]

Forward osmosis, a fairly novel technology, has received recent attention in Texas. The IWT has an ongoing project with CH2M Hill Companies Ltd to investigate the use of RO concentrate as a draw solution to recover wastewater using forward osmosis technology. According to the TWDB's project website, "The study will develop a framework to provide guidance to the industry in selecting the forward osmosis process as a component in an integrated system for simultaneous treatment of impaired water and brine from desalinated processes."[42]

An MD research project sponsored by the Desalination and Water Purification Research and Development Program at the U.S. Bureau of Reclamation tested a small commercial MD module and gathered data on the module's operations.[43]

WATER REUSE

As reliable, high-quality water becomes an ever scarcer resource, the reuse of treated wastewater for beneficial purposes has received notable consideration as a potentially effective way to supplement more traditional water supplies, and much effort has been invested in water reuse technologies. In recent years, water reuse has been applied in many sectors, from industry to agriculture, relieving a portion of water demand and proving to be a feasible and useful substitute water resource.

Water reuse is particularly attractive where available water supplies are already overcommitted and cannot meet expanding demand in a growing community. Secondary wastewater effluent can be of sufficient quality for some water reuse applications, such as agricultural irrigation and industrial cooling. More demanding applications, such as indirect reuse, require more advanced treatment for the removal of salts, organic compounds, and microorganisms.

Conventional Wastewater Treatment

Conventional secondary wastewater treatment removes suspended solids, dissolved organic matter, nutrients, and pathogenic microorganisms. The removal of these pollutants can meet the water quality requirements of secondary effluent and serve as pretreatment for more advanced water reuse applications. Important variations on conventional wastewater treatment include more advanced technologies for additional removal of nutrients. Conventional treatment processes include those technologies that remove dissolved organic matter, microorganisms, suspended solids, and some nutrients. Advanced treatment technologies achieve the same goals but use more advanced techniques that enhance nutrient removal and efficiency.

Conventional Activated Sludge Process. In the conventional activated sludge process, bacteria, growing either suspended in the water or attached to a solid phase (a membrane or other large surface area support), biologically remove suspended solids, dissolved organic matter, nutrients, and pathogenic microorganisms. Three methods are most commonly used in water reuse applications: suspended growth, attached growth, and hybrid processes.[44] In the suspended growth method, the microorganisms responsible for the treatment are maintained in suspension by appropriate mixing, and applications exist in which the reactor is operated either with or without aeration. In the attached growth method, or biofilm, the microorganisms are attached to an inert packing material such as rock, gravel, slag, or any of a wide range of plastic or other synthetic materials. The hybrid processes are mixtures of both suspended growth and attached growth processes.[45] Conventional treatment technologies have both advantages and disadvantages. Among the advantages are that the processes are familiar and well understood; they can be automated to a fairly high degree; and highly skilled operators may not necessarily be required except when nutrient removal is involved. The disadvantages are that these technologies may require larger physical facilities, such as clarifiers; they may be more susceptible to process upset; and effluent quality, especially total suspended-solids concentrations and turbidity, may be more difficult to control and thus may not meet the water quality requirements for reuse without special process augmentation.[46]

Conventional Activated Sludge Process with Nutrient Removal. The conventional activated sludge process may also integrate nutrient removal with the main biological activated sludge process, add chemicals to remove nutrients, or implement a process for the removal of a specific nutrient. Nitrogen and phosphorus removal are the two most common nutrient removal procedures in water reuse applications.[47]

Nitrogen removal is generally a biological process in which wastewater is treated in an aerobic-anoxic environment. Ammonia in influent is oxidized to nitrate and is then reduced to nitrogen gas, which bubbles out of the wastewater system. A variety of technologies can be used for nitrogen removal, including suspended growth processes, such as Modified Ludzack-Ettinger, step feed, sequencing batch reactors, and oxidation ditch; submerged attached growth processes, such as the Biofor and Biostyr processes; or fixed film packing, such as the Captor and Linpor and the Kaldnes processes.[48]

Phosphorus removal can involve either biological or chemical processes. Enhanced biological phosphorus removal (EBPR) is a biological process in which wastewater is treated first anaerobically, followed by an aerobic stage. Polyphosphate-accumulating organisms (bacteria that store granular phosphate) release polyphosphate during the anaerobic stage and accumulate polyphosphate during the aerobic stage. The polyphosphate-accumulating organisms are removed as sludge during the aerobic treatment stage, resulting in reduced phosphate concentrations. Example technologies for EBPR are the Phoredox and A^2O processes.[49]

Membrane Bioreactor Process. The membrane bioreactor (MBR) process combines biological treatment with a liquid-solid separation process that eliminates the need for secondary clarifiers. Two types are most commonly used in water reuse applications: submerged MBRs, where liquid-solid separation by the membranes may occur in the same basin where biological degradation occurs; and separated MBRs, where liquid-solid separation occurs as a separate process following activated sludge treatment. Advantages of the MBR process are a high-quality effluent with greater reuse potential, low suspended-solids concentrations and removal of large particles in product water with improved disinfection efficiency, a smaller footprint, and a potential reduction in sludge volume. Disadvantages include the need for pretreatment to avoid damaging and clogging membrane elements, greater consumption of energy for effective operation, and a potential for membrane fouling.[50]

Advanced Treatment for Particle Removal

For many water reuse applications, removal of residual particulate matter remaining after secondary biological treatment is required. Particulate matter contributes to turbidity, may be associated with undesirable chemical contaminants or pathogens, and may interfere with downstream processes such as reverse osmosis, adsorption, or disinfection. Particle removal following wastewater treatment can be accomplished by using conventional deep-bed filtration or dissolved air flotation. However, membrane filtration processes, such as ultrafiltration (UF) or microfiltration (MF), are likely to be more cost-effective.

Membrane filtration is a pressure-driven process that uses a semipermeable (porous) membrane to separate particulate matter from soluble components in the carrier fluid (e.g., water). The difference between MF and UF membranes is the size of the pores in the filter medium, which can vary from 0.005 to 2.0 micrometers. The use of membrane filtration in water reuse applications typically follows biological treatment. Most of the benefits of MF or UF are realized simultaneously when MBRs are used. Membrane filtration has several advantages over conventional filtration: reduced chemical use, smaller space requirements, reduced labor requirements because of easy automation, and chemical-free disinfection, including removal of virtually all protozoan cysts, oocysts, and helminthes ova. Disadvantages include a higher use of electrical power, possible need for pretreatment, and the need to manage fouling.

Removal of Dissolved Constituents with Membranes

The use of water typically leads to an increase in the concentration of dissolved contaminants, including salts; these must then be reduced in concentration before water reuse. Reduction in total dissolved solids is a reason that membrane technologies, described above in the section on desalination, are likely to be at the heart of a wastewater reuse facility.

Removal of Residual Trace Constituents

Although RO membranes are capable of removing the vast majority of compounds of concern in a water reuse scenario, trace constituents may remain that must be removed or destroyed by adsorption processes, such as ion exchange, granular activated carbon (GAC) adsorption, or oxidation. The purpose of adsorption in water reuse applications is to remove refractory organic constituents; residual inorganic constituents such as nitrogen, sulfides, and heavy metals; and odor compounds. In general, the process of adsorption can be described as removing substances that are in solution by accumulating them on a solid phase. The adsorption processes that are most commonly used in water reuse applications are fixed bed GAC column, expanded bed GAC, and mixed powdered activated carbon (PAC) contactor.[51] The treatment by GAC involves passing the water through a bed of activated carbon held in a reactor. In a fixed bed GAC column, the carbon is held in place with an underdrain system at the bottom of the column,[52] and the influent is applied to the top of the column. In an expanded bed column, the influent is introduced at the bottom and the carbon is allowed to expand.[53] PAC is generally added to the water to be treated and allowed to contact for a certain amount of time. The carbon is removed from the water by settling to the bottom of the contacting tank.

Ion exchange is an adsorption process whereby ions that are initially present on the adsorption medium (ion exchange resin) are swapped for ions in the feedwater. Ion exchange can remove common ionic constituents, such as Na^+ and Cl^-, and can soften water by removing calcium and magnesium ions (Ca^{2+} and Mg^{2+}). Ion exchange can also be used to remove specific constituents, such as barium, radium, arsenic perchlorate, chromate, and potentially other ions. The principle of ion exchange is the competition and replacement of mobile counterions for fixed counterions associated with fixed charged functional groups located on the exterior surface of the resin by electrostatic attraction. Ion exchange has several advantages: the running cost is very low; little energy is required; the regeneration chemicals are cheap; and if well maintained, the resin beds can last for many years before replacement is needed. Disadvantages include the possibilities of calcium sulfate fouling, iron fouling, adsorption of organic matter, organic contamination from the resin, bacteria contamination, and chlorine contamination.

Chemical oxidation may be required for odor control, hydrogen sulfide control, color removal, iron and manganese oxidation, control of biofilm growth and biofouling in treatment processes and distribution system components, and destruction of selected trace organic constituents. Many kinds of oxidants are used in chemical oxidation processes; listed in order of electrical potential, from high to low, they are ozone, hydrogen peroxide, chlorine dioxide, permanganate, and chlorine. The advantage of chemical oxidation is that it is very effective in the transformation and destruction of trace constituents. Disadvantages are the high cost of chemical addition and potential formation of toxic by-products.

Advanced oxidation processes (AOP) use multiple oxidizing agents (chemical or photocatalytic) in concert to enhance the formation of hydroxyl radicals (HO·), which react with the constituents in water. The purpose of AOP in water reuse applications is to destroy trace constituents that cannot be oxidized completely by a conventional oxidant such as endocrine disruptors. The principle of AOP involves the generation of HO·, which are strong oxidants capable of the complete oxidation of most organic compounds into carbon dioxide, water, and mineral acids. The three types of AOP used most commonly in water reuse applications are hydrogen peroxide and ultraviolet radiation (H_2O_2/UV), hydrogen peroxide and ozone (H_2O_2/O_3), and ozone and ultraviolet radiation (O_3/UV).[54] The advantage of AOP is that it is very effective for the transformation and destruction of trace constituents, often resulting in the complete mineralization of trace constituents. Disadvantages include the high cost of chemical addition and potential formation of toxic by-products.

Disinfection

Water reuse technologies must ultimately ensure that the finished water has an acceptably low number of potential pathogens and other microorganisms as defined by the required end use. Disinfection is the process used to achieve a given level of destruction or inactivation of pathogenic organisms. There are four categories of human enteric organisms found in treated water that potentially produce disease and thus must be destroyed in the disinfection process: bacteria, protozoan oocysts and cysts, helminthes, and viruses. The main disinfection technologies involve the use of chlorine, ultraviolet radiation, chlorine dioxide, ozone, and other chemicals.[55] Chlorine and UV disinfection are the two most commonly used treatments in water reuse applications.

Treatment Trains in Water Reuse

The type of treatment technologies used, the configuration of the treatment process, and the degree of reliability required for the treatment system are governed by the type of water reuse application. End uses such as indirect reuse (e.g., treatment followed by discharge to surface water or groundwater that subsequently serves as the source for a potable water treatment facility) require extensive treatment. A flow diagram of such a demanding treatment process including many of the technologies discussed above is shown in Figure 10.2.

Reuse for agricultural or landscaping purposes can be accommodated by a less intense level of treatment. The constituents in reclaimed water of concern for agricultural reuse are salinity, sodium, trace elements, chlorine residuals, and nutrients. The concentration of these constituents in secondary effluent is

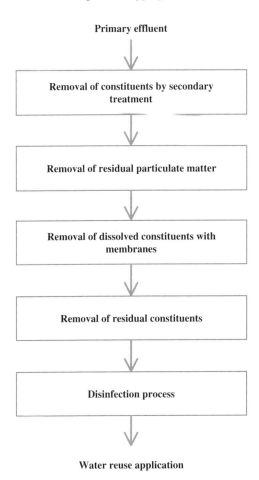

Figure 10.2 *Possible treatment processes for a high-quality end use*

generally under the tolerance level, so no specific treatment is required. However, in some cases where special crops are involved, additional requirements may be imposed for the removal of certain constituents. Figure 10.3 shows a typical configuration of a treatment process for agricultural reuse, without any special treatment requirements.

Water Reuse Technologies in Texas

Water reuse technologies are already a significant part of the Texas water supply. The state water plan indicates that 14 of the state's 16 planning regions include water reuse, either direct or indirect or both, as a water management strategy.[56] Water reuse applications in the state seem to be correlated with two factors: arid climate and municipal size. Where these two factors converge—in a big city in a dry part of the state—quantities involved in the reuse of municipal water are impressive. In 1998, Bexar County, which includes the city of San Antonio, was

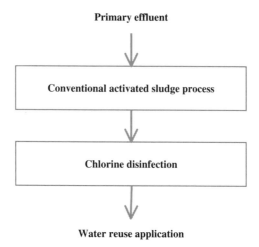

Primary effluent

Conventional activated sludge process

Chlorine disinfection

Water reuse application

Figure 10.3 *Typical treatment process flow diagram for agricultural reuse*

reusing water at a rate of 37.8 MGD. Other major water-reusing counties (and the quantity of water each reused in 1998 in MGD) include Collin (29.0), Potter (13.9), Midland (10.5), and Travis (9.4).[57] These counties all include or are associated with cities that have medium to large populations.

The Dallas–Fort Worth Metroplex currently recycles approximately 5% of its wastewater, primarily for public irrigation. Wastewater reuse systems utilizing conventional wastewater treatment followed by reverse osmosis are currently being used to augment industrial water supplies in the Texas Lower Rio Grande Valley cities of McAllen and Harlingen. Approximately 5% of municipal wastewater in this area is reused, with 10,687 acre-feet per year for direct reuse and 2,240 acre-feet per year for indirect reuse.

As of 1998, the western half of the state, which receives less rainfall, had a greater proportion of counties that were involved in water reuse. The reuse of industrial water is common in western Texas; most of the top-ranked counties in the state for industrial reuse are in this region. Potter, Lubbock, El Paso, and Ector Counties reuse, respectively, 10.67, 4.03, 3.70, and 2.09 MGD of their industrial water. Nueces County, on the Texas Gulf Coast and part of the Corpus-Christi metropolitan area, was also reported to reuse 3.09 MGD of its industrial water.[58]

Many other interesting and notable examples of water reuse projects exist in Texas. In El Paso, the Fred Hervey Water Reclamation Plant provides highly treated wastewater for a variety of nonpotable uses.[59] The city of Austin uses treated wastewater to fill lakes and ponds on the municipal golf course. Treated wastewater also irrigates the grass outside the famous Texas independence site, the Alamo, in San Antonio.[60] On South Padre Island, the wastewater treatment plant discharges effluent into a wetland-marsh area, around which a nature trail has been constructed.[61]

As is the case with desalination, the IWT group of the TWDB is active in promoting, financing, and facilitating water reuse applications around the state. The IWT's webpage on water reuse contains links to current projects, completed reports, and a database of water reuse projects around the state. The database lists information on water reuse projects, including the participating entities, the purpose of each project, and the quantities of water involved. Project entries in the database can be browsed by the project's size (i.e., quantity of water reused), year, or county.[62]

HARVESTED WATER

Another means of obtaining water is through water harvesting. Sources of harvested water include rainwater, humidity, and water that is produced as a by-product from the use of fuel cells.

Rainwater Harvesting

Rainwater can be captured, diverted, and stored to provide inexpensive water for residential, commercial, and industrial purposes. Harvested rainwater can be readily used for irrigation and nonpotable in-house applications, and disinfection techniques can be used to make it potable. People have harvested rainwater since antiquity, and it is currently done in dry and rural areas where a central utility for water supply is not available and water sources are scarce. Some towns also use this method of obtaining water in conjunction with the local supply system.

A rainwater harvesting system typically consists of subsystems for catchment, storage, and distribution. In a roof catchment system, gutters and downspouts are used to capture water that falls on the collection area and direct it into the storage tank. A similar system, but on a larger scale, is used for industrial, commercial, and community rainwater harvesting. Because of economies of scale, industrial and commercial systems typically are less expensive than residential per unit of water harvested. On the other hand, large storage systems are costly, and treatment of the water is required before use. Ground and rock catchment surfaces are used in these large-scale systems, and water is collected from a much larger area than with an average roof catchment.[63]

A major difference from desalination and reuse technologies is that rainwater harvesting is associated with relatively high-quality raw water, which can be further improved with simple treatment technologies. Rainwater has potentially advantageous characteristics, such as softness, and a slightly acidic pH. Furthermore, rainwater is free of salts, minerals, and other contaminants. Nevertheless, rainwater is exposed to several sources of contamination, both natural and artificial. One method for improving the quality of rainwater in roof catchment systems is the use of leaf screens to prevent leaves, insects, and other coarse material from entering the water collection system. Finer and dissolved material, however, such as natural and chemical contaminants that have been deposited on the surface of a roof (e.g., animal feces, pollen, or pesticides), have been washed from the atmosphere (e.g., aerosols), or have dissolved in the water from the roof material (e.g., polymers), are not filtered out by the leaf screens. First-flush diverters are used to inhibit these finer substances from entering the water storage tank. Drip irrigation systems also have filtering units for protection against clogging of the irrigation emitters.

To make harvested rainwater potable, a roof washer is used ahead of the storage tank to filter small particles, and a treatment and disinfection system is installed before the tap. This typically consists of two filters in series and a UV disinfection unit. Because rainwater has low concentrations of contaminants compared with other water sources, filtration techniques are easier to use for treatment. Membrane filtration removes dissolved solids from the collected water. Ozone and chlorine treatments can be used for disinfection but are less common. Disinfection also may be implemented to protect from contaminants such as atmospheric particles, combustion by-products, chemicals, and radioactive isotopes.

Energy requirements for a rainwater harvesting system are significantly lower than those for humidity harvesting, as the rainwater system may work solely on gravitational flow or require just a few pumps.

Humidity Harvesting

Humidity harvesting is a method of condensing ambient humidity in the air for reclaiming water. Two types of humidity-harvesting systems are prevalent: those designed to deliver water by means of condensing air, and those that are essentially air-conditioning equipment that produces condensate as a by-product. These technologies are similar in principle and enable the production of water almost anywhere, because the source—air—is abundant, and the units can be mobile. It is especially advantageous in low-precipitation regions around the equator because of the relatively higher humidity levels. During low-rainfall periods, it could provide a viable water source for remote and rural areas, as air humidity typically peaks at this time of year. Cooling of the air causes part of the ambient humidity to condense into liquid water. Case-specific thermodynamic analysis is necessary for calculating the amount of water that can be obtained by this method, but in a typical hot and humid day, a household air-conditioning unit can produce more than 10 liters of pure water.[64] Industrial and commercial facilities generate thousands of liters per day, but the exact amount greatly depends on the size and type of industry, local weather conditions, and the type of air-conditioning systems used. This is a source of relatively pure, free-of-charge water that potentially could be used for both nonpotable and potable applications. Current applications include mobile units for temporary water production after natural disasters and for military purposes, as well as air-conditioning systems.

The main disadvantage of this technology is the volume ratio between vapor and liquid water: 1,600 liters of air with 100% relative humidity correspond to just 1 liter of liquid water. The amount of air that needs to be condensed for the production of a prespecified amount of water depends mainly on the relative humidity of the air and the operating efficiency of the humidity condenser. For example, a condenser operating at 100% efficiency will have to process 3,200 liters of air with 50% relative humidity to produce just 1 liter of liquid water. Nevertheless, condensate is high-quality water that can be found practically everywhere, as it is drawn out of the air. Currently, however, the technology is not developed to a point that would make it cost-effective for wide application.

Water Fabrication: Production from Fuel Cells

Fuel cells produce energy through an electrochemical reaction using hydrogen (or potentially alcohols) as a fuel source, in pure form or from hydrocarbons in conventional fuels, and oxygen as an oxidant. Fuel cells have attracted a lot of attention in the past few years as an alternative energy source because of several of their promising characteristics, including clean emissions (the by-products are heat, water, and carbon dioxide), no combustion (the process is quiet and involves no moving parts), abundant fuel in the form of hydrogen and oxygen, durability, and efficiency.[65]

The initial electrochemical reaction produces protons and electrons, which are forced to follow different paths. In polymer electron membrane (PEM) fuel cells, the membrane is permeable only to protons. Electrons follow an external circuit to the cathode, thus creating electrical current, which is used for electricity. Water is produced as a by-product based on the following electrochemical reactions:

$$H_2 \rightarrow 2H^+ + 2e^- \quad \text{(anode)}$$

$$\frac{1}{2}O_2 + 2H^+ + 2e^- \rightarrow H_2O \quad \text{(cathode)}$$

Fuel cells have numerous potential and existing uses. They are being employed for backup and auxiliary power, portable power, transportation, large and small distributed generation, and military and space purposes. Their applications can be mobile (e.g., vehicles, spacecraft, or portable batteries) as well as stationary (e.g., use in buildings or remote stations). The level of technological maturity and marketability varies among different applications. For example, for the production of electricity in vehicles and mobile batteries, fuel cells are in the experimental and prototype stages, but they already have been implemented as power generators in spacecraft.

The water that is generated as a by-product in the energy production process is free of contaminants and thus potentially can be used for potable and nonpotable applications. In the case of spacecraft, the water produced from fuel cells is used as potable water. Fuel cells for the production of electricity are currently in operation in hospitals, research centers, and universities, and the water that is generated possibly could serve as an additional water source for use within these buildings.

Approximately 1 liter of water per kilowatt-hour is produced as a by-product of the fuel cell process. A large portion of this volume, about 70%, is used to keep the fuel cell membranes hydrated. Typical energy production efficiency of fuel cells is approximately 30 kWh/kg of hydrogen, while 9 kg of water are produced per kg of hydrogen used. Thus, for every kWh produced from a fuel cell, 0.21 kg of water is used for keeping the membranes hydrated, and 0.09 kg of pure water is disposed of as a by-product.

Harvested Water in Texas

Of the three technologies discussed in this section on harvested water, rainwater harvesting far and away has received the most attention from state groups and water conservation advocates. Not surprisingly, the IWT group of the TWDB hosts an informational service website with information about its past and present rainwater-harvesting activities, as it does for desalination and water reuse.[66] The IWT rainwater-harvesting page includes links to a rainwater-harvesting system size calculator, information about the Texas Rain Catcher Award, and a statewide map of estimated rainwater yields (see Figure 10.4).

Texas A&M University also hosts an informational service website, with demonstrations, publications, training opportunities, rainwater-harvesting system suppliers, and an inventory of other useful information for those interested in

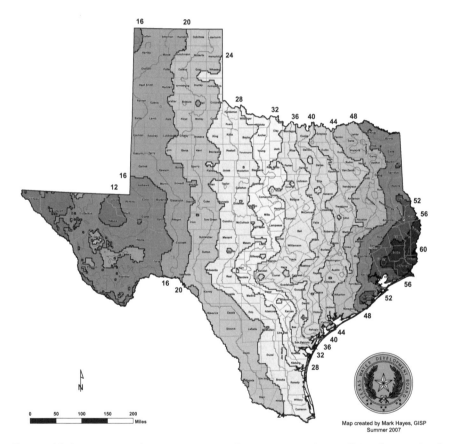

Figure 10.4 *Rainwater harvesting: map of average annual runoff (in thousands of gallons) from a typical 2,000-square-foot roof*

Source: TWDB, Rainwater Harvesting: Average Annual Runoff (in Thousands of Gallons) from a Typical 2,000 sq. ft. Roof (Austin, TX: TWDB, 2007), http://www.twdb.state.tx.us/iwt/rainwater/faq.asp#title-05 (accessed June 2, 2010).

harvesting rainwater and conserving water.[67] This website has contributions from groups in Texas as well as from other states in the Southwest.

ECONOMIC AND OTHER CONSIDERATIONS

The considerations for implementation of advanced water technologies are as diverse as they are significant. Each technology discussed in this chapter has a unique set of capital and operational requirements. But capital and operational expenses are but two of the many items that must be considered during a water project's planning stage. For example, the riskiness of a novel technology in a full-scale setting must be weighed against the technology's theoretical, or pilot-scale, promises of efficiency improvement. A variety of other site-specific characteristics also must be considered in the context of each technology, such as the accessibility of concentrate disposal sites in the case of desalination projects or the abundance of rainfall and a workable collection surface in the case of rainwater-harvesting systems. Because of the heterogeneity among the different technologies' site-specific costs and considerations, comparing two or more in any general way is often a difficult task. Nevertheless, to attempt a general discussion that compares costs across all technologies is the intent of this section. A useful place to start is to identify any commonalities shared by all the technologies in this chapter, so this section begins with a comparison of power consumption. Reported costs for advanced-technology facilities in the state as well as across the world then are discussed for selected technologies.

A common cost item shared by all the technologies is the power cost for the energy associated with operation. As the price of energy varies from location to location, a useful indicator for power costs of a variety of available technologies is their energy requirements. Table 10.1 lists the technologies discussed in this

Table 10.1 *Energy consumption for the production of 1 cubic meter of water with the various technologies*

Technology	Energy consumption (kWh/m³)
Wastewater reuse: agricultural	3.2
Wastewater reuse: industrial or indirect potable	10.2
Desalination: multistage flash	51.7
Desalination: multiple effect evaporation	45
Desalination: mechanical vapor compression	10–11
Desalination: geothermal	Very low: 2 pumps and a blower
Desalination: solar	0.3
Desalination: reverse osmosis	6.7
Desalination: electrodialysis	7–10
Desalination: forward osmosis	20–30
Rainwater harvesting	Very low: 1 pump
Humidity harvesting	2,575
Fuel cell[a]	−3,333

[a] Negative energy consumption values correspond to energy output by the technology.

chapter along with the corresponding energy requirement for the production of 1 cubic meter of water. The more energy-consumptive technologies include MSF and MEE, whereas RO, ED, and MVC are midrange energy consumers. The fuel cell, however, is actually an energy-generating device, and therefore this technology has a negative level of energy consumption.

Capital costs of each technology depend on local conditions such as the availability of materials and the quality of the source water. A survey of desalination facilities in the state was conducted in conjunction with the establishment of the state's desalination database. The survey's findings included capital cost values for desalination facilities in the state. These capital costs ranged from $0.05 to $1.64 per 1,000 gallons (not adjusted for inflation and assuming a 20-year lifespan).[68] In a separate research project, Sturdivant and colleagues conducted an extensive case study of a single 7.5 MGD brackish groundwater desalination facility near Brownsville. In addition to capital costs, the researchers included in their total cost calculation the facility's continued costs, such as electricity, labor, and the periodic replacement of facility components (e.g., RO membranes and pumps). Sturdivant et al. reported a total calculated cost of $2.34 per 1,000 gallons for combined construction and continued costs.[69]

Capital costs for very large MSF plants in the Arabian Peninsula are approximately $0.49 per 1,000 gallons; very large MEE plants in the region require approximately $0.6 per 1,000 gallons in capital costs (assuming a lifespan of 20 years).[70] Some of the technologies are very cheap to install, as long as adequate environmental conditions exist. For example, on the island of Nisyros in Greece, a small geothermal desalination unit capable of producing 75,000 gallons per day cost approximately $780,000 (or €600,000 at 2007 exchange rate). A 1992 study estimated the capital costs of a solar humidification unit as between $2 and $6.4 per 1,000 gallons.[71] The capital costs ($/gallons per day [GPD]) of water reuse projects decrease with the size of the system. In New York, the capital costs of water reuse projects range from $50/GPD for a 10,000 GPD plant to $15/GPD for a 500,000 GPD plant.[72]

The *Texas Manual on Rainwater Harvesting* states that the single largest contributor to a rainwater-harvesting system is the storage tank, which can vary by size and material.[73] Storage tank costs range from $0.50 to $4.00 per gallon. Beyond the storage tank, other materials include gutters, roof washers, pumps and pressure tanks, and filtration and disinfection equipment. All components of a rainwater-harvesting system vary in cost depending on size, materials, and whether the components are installed by professionals. In Texas, financial incentives and tax exemptions encourage the implementation of these systems. Many local entities provide other incentives, such as rebate programs and rain barrel distribution events, to further promote the adoption of rainwater-harvesting systems.

CONCLUSIONS

The IWT group and the many research programs at the state's universities are looking to, and evaluating, the future of advanced water treatment technologies in

Texas. Many locations around the state already have initiated or completed water projects that implement advanced water treatment technologies. These efforts are motivated by growing awareness of the disparity between water supplies and demands in parts of the state.

Many of the more tried and true advanced technologies have been implemented in the state, such as brackish groundwater RO, water reuse for nonpotable purposes, and rainwater harvesting. Highlights of the state's initiative to develop advanced-technology water projects include the world's largest inland desalination facility in El Paso and the two seawater desalination pilot projects in Brownsville and South Padre Island. These projects and many others were funded in part by the TWDB, which makes results from research projects, such as the desalination pilots, available to water utilities and managers in Texas, as well as the rest of the world via the TWDB website. State funding of technologically novel water projects, pilot projects in particular, may be justified by their benefits—specifically, an enhanced and more informed knowledge base—which accrue to all water stakeholders across the state.

In addition to RO, rainwater harvesting, and water reuse, activities around the state's research universities are investigating and implementing technologies on the cutting edge of water engineering. A striking example is the development of AdVE and the technology's subsequent pilot testing in Laredo. In addition to water treatment, research initiatives around the state are studying the coupled system of energy and water production. The Global Petroleum Research Institute is investigating the recycling of brackish or oily water recovered during oil and gas exploration.[74] The Wind Science and Engineering Research Center at Texas Tech University and General Electric have partnered in an effort to study wind power with brackish water desalination in west Texas.[75] Recently, researchers from the Marine Engineering and Technology department at Texas A&M University and Minnesota-based INRI have tested a wave-driven pump capable of providing energy to a seawater desalination facility.[76] The study of water energy systems, especially in conjunction with renewable energy, will become increasingly important as water becomes scarcer and sensitivity to global climate change increases.

Perhaps the most informative aspect to the state's involvement with advanced water technology is the way each region's unique set of circumstances dictates its interest in and application of advanced water treatment technologies. In many metropolitan areas, water reuse has emerged as a common aid to water scarcity, filling water demand niches such as landscape and golf course irrigation and natural habitat restoration. Areas along the Rio Grande, such as El Paso, Laredo, and Brownsville, which have been primarily dependent on the Rio Grande for their water supplies, are now looking to brackish aquifers and desalination technologies to augment current water supplies and hedge against rises in demand driven by population growth. Along the Gulf Coast, where the prospect of seawater desalination holds the most potential, the state is conducting its second pilot project of the technology. In all cases across the state, no single water supply technology is in place or being promoted. Each region is taking stock of available resources and selecting paths to manage water scarcity. Many advanced technologies are neither cheap nor free of risk. Effective water management may be characterized by the thoughtful management of existing water supplies and measured implementation of advanced water technologies.

NOTES

1. H.T. El-Dessouky, H.M. Ettounney, and Y. Al-Roumi, Multi-Stage Flash Desalination: Present and Future Outlook, *Chemical Engineering Journal* 73, no. 2 (1999): 173–90.

2. Ibid.

3. Ibid.

4. Ibid.

5. N.M. Wade, Distillation Plant Development and Cost Update, *Desalination* 136 (2001): 3–13.

6. H.T. El-Dessouky, H.M. Ettounney, and F. Mandani, Performance of Parallel Feed Multiple Effect Evaporation System for Seawater Desalination, *Applied Thermal Engineering* 20, no. 17 (2001): 1679–1706.

7. Wade, Distillation Plant Development.

8. H. Ettouney, H. El-Dessouky, Y. Al-Roumi, Analysis of Mechanical Vapour Compression Desalination Process, *International Journal of Energy Research* 23 (1999): 431–51.

9. J.M. Veza, Mechanical Vapor Compression Desalination Plants: A Case Study, *Desalination* 101 (1995): 1–10.

10. K. Bourouni, J.C. Deronzier, and L. Tadrist, Experimentation and Modeling of an Innovative Geothermal Desalination Unit, *Desalination* 125 (1999): 147–53.

11. R.J. Erdlac, Geopowering Texas: A Report to the Texas State Energy Conservation Office on Developing the Geothermal Energy Resources of Texas, 2007, http://www.seco.cpa.state.tx.us/zzz_re/re_geopowering2007.pdf (accessed June 2, 2010)

12. S. Kalogirou, Survey of Solar Desalination Systems and System Selection, *Energy* 22, no. 1 (1997): 69–81.

13. K. Schwarzer, M.E. Vieira, C. Faber, and C. Muller, Solar Thermal Desalination System with Heat Recovery Desalination, *Desalination* 137 (2001): 23–39.

14. E. Chafik, A New Seawater Desalination Process Using Solar Energy, *Desalination* 153 (2002): 25–37.

15. Kalogirou, Survey of Solar Desalination Systems.

16. H.J. Krishna, Introduction to Desalination Technologies, in *The Future of Desalination in Texas*, ed. Jorge A. Arroyo, vol. 2, *Technical Papers, Case Studies and Desalination Technology Resources*, ch. 2, Desalination Technology (Austin, TX: Texas Water Development Board, 2004), available at http://www.twdb.state.tx.us/iwt/desal/docs/Volume2Main.asp (accessed May 26, 2010).

17. T. Schnettler, Desalination Method to Benefit Laredo with Clean Water, Department of Chemical Engineering, Texas A&M University, March 23, 2009, http://www.che.tamu.edu/department/desalination-method-to-benefit-laredo-with-clean-water (accessed September 1, 2009).

18. R.E. Foster, W. Amos, and S. Eby, Ten Years of Solar Distillation Application along the U.S.-Mexico Border (presented at the Solar World Congress, International Solar Energy Society, Orlando, FL, August 11, 2005).

19. The El Paso Solar Energy Association (EPSEA) maintains a website at http://www.epsea.org (accessed September 13, 2010).

20. B. Lesikar, Task 6: Environment, Ecology and Water Quality Protection. Memorandum of Agreement (College Station, TX: Texas Water Resources Institute, 2006), http://riogrande.tamu.edu/deliverables-view/2006-07/168 (accessed November 25, 2009; site now discontinued).

21. J.P. Nicot, S. Walden, L. Greenlee, and J. Els, *A Desalination Database for Texas*, 2006, http://www.beg.utexas.edu/environqlty/desalination/Final%20Report_R1_1.pdf (accessed June 2, 2010).

22. Wade, Distillation Plant Development.

23. The Dow Chemical Company, World's Largest Desalination Plant Running with Filmtec™ Membranes Is Now Fully Commissioned, news release February 9, 2006, The Dow Chemical Company, http://www.wateronline.com/article.mvc/Largest-Desalination-Plant-Utilizing-Filmtec-0001 (accessed September 13, 2010).

24. V.V. Slesarenko, Electrodialysis and Reverse Osmosis Membrane Plants at Power Stations, *Desalination* 158 (2003): 303–11.

25. T.Y. Cath, A.E. Childress, and M. Elimelech, Forward Osmosis: Principles, Applications, and Recent Developments, *Journal of Membrane Sciences* 281 (2006): 70–87.

26. J.R. McCutcheon, R.L. McGinnis, and M. Elimelech, A Novel Ammonia–Carbon Dioxide Forward (Direct) Osmosis Desalination Process, *Desalination* 17 (2005): 1–11.

27. S.T. Hsu, K.T. Cheng, and J.S. Chiou, Seawater Desalination by Direct Contact Membrane Distillation. *Desalination* 143 (2002): 279–87; B. van der Bruggen and C. Vandecasteele, Distillation vs Membrane Filtration: Overview of Process Evolutions in Seawater Desalination, *Desalination* 143 (2002): 207–18; Cath et al., Forward Osmosis.

28. M. Mulder, *Basic Principles of Membrane Technology* (Dordrecht, The Netherlands: Kluwer Academic Publishers, 1997).

29. Cath et al., Forward Osmosis.

30. L. Song, J.Y. Hu, S.L Ong, W.J. Ng, M. Elimelech, and M. Wilf, Performance Limitation of the Full-Scale Reverse Osmosis Process, *Journal of Membrane Science* 214 (2003): 239–44.

31. Texas Water Development Board (TWDB), *Water for Texas 2007*, Texas State Water Plan, Document No. GP-8-1 (Austin, TX: TWDB, 2007).

32. NRS Consulting Engineers (NRS), *Final Pilot Study Report: Texas Seawater Desalination Demonstration Project* (Austin, TX: Texas Water Development Board, 2008).

33. Nicot et al., *Desalination Database*.

34. TWDB, Desalination Database, 2005, http://www.twdb.state.tx.us/iwtdesaldb/dbStart.aspx (accessed November 25, 2009).

35. El Paso Water Utilities, Water (information webpage), 2007, http://www.epwu.org/water/desal_info.html (accessed November 25, 2009).

36. NRS, Desalination Process (website portal for information on innovations in desalination in Texas), 2009, http://www.desal.org (accessed November 25, 2009).

37. NRS, *Guidance Manual for Brackish Groundwater Desalination in Texas* (Austin, TX: Texas Water Development Board, 2008).

38. TWDB, *The Future of Desalination in Texas: Biennial Report on Seawater Desalination* (Austin, TX: TWDB, 2008).

39. TWDB, Innovative Water Technologies: Desalination, http://www.twdb.state.tx.us/iwt/desal.asp (accessed November 25, 2009).

40. Brazos River Authority website, http://www.brazos.org (accessed November 25, 2009).

41. Nicot et al., *Desalination Database*.

42. TWDB, Forward Osmosis/Reverse Osmosis: Project Brief: An Assessment of Osmotic Mechanisms Pairing Desalination Concentrate and Wastewater Treatment, 2009, http://www.twdb.state.tx.us/iwt/desal/studies/other/ch2mhill/docs/May09.pdf (accessed June 2, 2010).

43. J. Walton, L. Huanmin, C. Turner, S. Solis, and H. Hein, *Solid and Waste Heat Desalination by Membrane Distillation*, Desalination and Water Purification Research and Development Program Report No. 81 (Denver, CO: U.S. Department of the Interior, Bureau of Reclamation, 2004).

44. T. Asano, F.L. Burton, H.L. Leverenz, R. Tsuchihashi, and G. Tchobanoglous, *Water Reuse: Issues, Technologies, and Applications* (New York: McGraw-Hill, 2006).

45. Metcalf & Eddy, Inc., G. Tchobanoglous, F.L. Burton, and H.D. Stensel, *Wastewater Engineering: Treatment and Reuse*, 4th ed. (New York: McGraw-Hill, 2003).

46. Water Environmental Federation (WEF), *Design of Municipal Wastewater Treatment Plants, Manual of Practice 8* (Alexandria, VA: Water Environmental Federation, 1998).

47. Asano et al., *Water Reuse*.

48. Metcalf & Eddy, Inc., et al., *Wastewater Engineering*.

49. Ibid.

50. WEF, *Design of Wastewater Treatment Plants*.

51. Asano et al., *Water Reuse*.

52. Metcalf & Eddy, Inc., et al., *Wastewater Engineering*.

53. Ibid.

54. Ibid.

55. Asano et al., *Water Reuse*.

56. TWDB, *Water for Texas 2007*.

57. H.J. Krishna, Water Reuse in Texas, TWDB, http://www.twdb.state.tx.us/assistance/conservation/municipal/Reuse/ReuseArticle.asp (accessed October 18, 2009).

58. Ibid.

59. El Paso Water Utilities, Wastewater Treatment: Northeast–Fred Hervey Plant, 2007, http://www.epwu.org/wastewater/fred_hervey_reclaimation.html (accessed November 25, 2009).

60. R.J. Huston, Texas Water Rights and Wastewater Reuse, *Confluence* (Texas Water Conservation Association newsletter), 3rd qtr. 2006.

61. J.S. Jacob, D.W. Moulton, and R.A. Lopez, Texas Coastal Wetlands Guidebook, 2003, http://www.texaswetlands.org/lowercoast.htm (accessed November 25, 2009).

62. TWDB, Innovative Water Technologies: Water Reuse, http://www.twdb.state.tx.us/iwt/reuse.asp (accessed November 25, 2009).

63. J. Gould and E. Nissen-Petersen, *Rainwater Catchment Systems for Domestic Supply* (London: Intermediate Technology Publications, Ltd, 1999).

64. H.J. Krishna, *The Texas Manual on Rainwater Harvesting* (Austin, TX: TWDB, 2005).

65. N. Sammes, ed., *Fuel Cell Technology: Reaching towards Commercialization* (Storrs, CT: Springer-Verlag London Limited, 2006).

66. TWDB, Innovative Water Technologies: Rainwater Harvesting, http://www.twdb.state.tx.us/iwt/Rainwater.asp (accessed November 25, 2009).

67. Texas A&M University, AgriLife Extension, Rainwater Harvesting, http://rainwaterharvesting.tamu.edu (accessed November 25, 2009).

68. Nicot et al., *Desalination Database*.

69. A.W. Sturdivant, C.S. Rogers, M.E. Rister, R.D. Lacewell, J.W. "Bill" Norris, J. Leal, J.A. Garza, and J. Adams, Economic Costs of Desalination in South Texas: A Case Study, *Journal of Contemporary Water Research & Education* 137 (2007): 21–39.

70. K. Wangnick, *White Paper: Present Status of Thermal Seawater Desalination Techniques*, 2004, http://www.idswater.com/Common/Paper/Paper_51/Present%20Status%20of%20Thermal%20Seawater%20Desalination.htm (accessed October 1, 2009).

71. D. Hoffman, Application of Solar Energy for Large-Scale Seawater Desalination, *Desalination* 89, no. 2 (1992): 115–83.

72. E. Clerico, testimony, The Future of Water Reuse in America, Hearing of U.S. House Subcommittee on Energy and Environment, 2007, http://democrats.science.house.gov/Media/File/Commdocs/hearings/2007/energy/30oct/Clerico_testimony.pdf (accessed June 2, 2010).

73. Krishna, *Texas Manual on Rainwater Harvesting*.

74. Global Petroleum Research Institute, Conversion of Oil Field Produced Brine to Fresh Water, Texas A&M University, 2003, http://www.pe.tamu.edu/gpri-new/home/Conversion Brine.htm (accessed March 1, 2009).

75. Wind Science and Engineering Research Center, Overview: Wind Energy, Texas Tech University, 2009, http://www.wind.ttu.edu/WindEnergy/Overview.php (accessed March 1, 2009).

76. Red Orbit, Renewable Energy Ocean Wave Technology Validated by Texas A&M University's Report on Independent Natural Resources, Inc.'s SEADOG(R) Pump, news release, May 20, 2008, http://www.redorbit.com/news/business/1394737/renewable_energy_ocean_wave_technology_validated_by_texas_am_universitys/index.html (accessed March 1, 2009).

CHAPTER 11

Water Management Guidance from Texas

Ronald C. Griffin

*I*t is difficult to review the progress of any government's water policy without feeling a sense of urgency and a need for reform. There are always policy improvements to be made in a world where more and more feet add to the human footprint, steadily thinning all water resources. Fortunately, some of Texas's experiments in water policy have worked out relatively well. Maybe these can be role models for other states and countries. Some of Texas's unique efforts to steward its water have had poor results. Perhaps these can be pinpointed and avoided. Myriad results lie somewhere in between. Let's consider the fundamental lessons provided by the prior chapters first.

CENTRAL LESSONS

Several chapters in this book support a finding that this physically diverse state has taken a diverse policy path (Chapters 2–9). Among the lessons from Texas are nineteenth-century changes. What began as a nation with a Spanish heritage in water law moved quickly to an eastern U.S. system in water law, based on English common law. Different directions ultimately were taken for surface water and groundwater (Chapter 3), and the hydrologic interconnectedness of these resources has yet to be matched by integrated management (Chapters 3 and 5). What are the changes, status, and standing issues within these two bodies of law?

Surface Water

For surface water, the English common law transition meant that Texas adopted the riparian doctrine (Chapter 3), with its attendant requirement that every water use be reasonable in relation to other basin uses, but this requirement was unclear, unquantified, and ultimately unsuitable for semiarid regions. So more than a century ago, Texas transitioned again, this time to a system of quantified water rights, as befits a water-scarce land. Annual fluctuations in flow were handled by seniority, with historically prior users having more senior claims. Because they were no longer bound to parcels of land, surface-water rights have long been theoretically transferable, even though the administrative support system (e.g., metering, enforcement) was underdeveloped. These water rights can be sliced so that the owner can keep some and transact some (or all). The divisibility of rights implies that junior standing or nonownership is an escapable condition for everyone, as long as newcomers have the cash to back up their wishes (Chapter 4). The fact that those with historically prior, lower-valued uses are getting this money makes them willing parties in economic growth—a very good thing in light of the weak alternatives for accomplishing water reallocation in an advancing economy.

For several decades, both preexisting Spanish/Mexican rights and riparian rights were honored alongside the newer appropriative rights, but the burdens of this confusion eventually were escaped by converting all surface-water claims into the appropriative style, commencing in the 1960s (Chapter 3). This overarching system is still in place in most of the state. For the lower Rio Grande, a distinct 1956–1970 court case condensed all surface-water claims into a single nonseniority system. This region's rights are also transferable among users. Thus, surface-water rights can be bought or sold throughout the entire state.

Given the above history, it is apparent that rising water scarcity fostered policy advance in the case of surface water. Today's frontline surface-water issues are interbasin transfers (Chapter 3), reserving environmental water for estuaries and instream flow (Chapters 6 and 7), and a population-induced demand for still more impoundments and infrastructure (Chapter 2). These are reviewed in more detail later.

Groundwater

In its early history, Texas also selected an English-rooted legal code for groundwater—the rule of capture—but in this case, meaningful evolution of water law did not occur (Chapter 3). It appears that progress was hampered by the "invisibility" of this resource and a tradition-ingrained viewpoint that groundwater is the unconstrained property of the overlying landowner. Invisibility allowed sufficient deniability about whether others were affected by pumping, even though they generally are, and tradition favored landowning pumpers in disputes. The Texas judiciary solidified this perspective through continued court rulings, and the state legislature did not initiate a new course. Only in the extraordinary case of the Edwards Aquifer, where a federal authority mandated that Texas do "something," did the legislature act to sever water rights from land rights and establish quantified groundwater rights for the first time, in 1993, for this single aquifer (Chapter 5). Even in the case of the new Edwards doctrine, however, Texas is having a hard time mustering the political will to stand by the pumping limits it was forced to adopt in 1993. The urge to enable popular visions of economic development is too great.

Outside of the Edwards Aquifer, the pressures of growth have been addressed by both groundwater mining and land-based transfers of groundwater access (Chapter 4). Conservation motives are very weak with a rule of capture, so Texas has experienced depletion (Chapter 9), and in some places depletion also has induced land subsidence or saltwater intrusion. Pumping lifts are steadily rising, with negative implications for all users, not just the biggest pumpers. Groundwater depletion is especially notable for the Ogallala Aquifer, where it has major implications for the very nature of that region's future economy (Chapter 9). Whereas depletion is normal human behavior with all exhaustible resources, there are reasons to believe that it continues to be too rapid for Texas groundwater.

Today's number-one issue for groundwater is the invigorated application of groundwater districts, which are trying to step into the management vacuum created by the rule of capture. The idea is to shift some decisionmaking responsibility to locally elected authorities with the ability to regulate or tax pumping, particularly when confronted by proposals to export water to distant areas (Chapters 3 and 9). A second, ongoing issue is difficulty in staying the course with Edwards Aquifer reform, and thereby granting firm protection for environmental services and downstream surface-water rights (Chapters 4 and 5). These two issues are considered below.

ADDITIONAL EXPERIENCES

If the last quarter century is a good gauge, change is now an integral part of Texas water policy. Yet the surface-water rights system possesses stable foundations in that the central doctrines are well established. On the groundwater side, a grand experiment concerning the adequacy of groundwater districts has been initiated. What individual messages can be extracted from the many experiences generated here?

Managing Water across Borders

In-state water management has been assisted when clarity exists over Texas's share of multijurisdiction water (Chapter 8). In such cases, state and local water authorities can know what they have to work with. Individuals can more confidently make investments, such as in irrigation equipment, and commitments to production activities. If border water agreements are not established or are poorly maintained, uncertainty will exist in the minds of water users and managers. Weather always injects unavoidable uncertainty into surface-water supply, so additional uncertainty arising from misperforming policy is not desirable.

Both interstate compacts and international treaties have contributed to stability, though unequally. The best of these arrangements have two characteristics: they specify divisions of water that anticipate dry times as well as normal periods, and they are enforceable. The U.S.-Mexico treaty has been a very positive force in water management, yet loose interpretation of subjective language such as "extraordinary drought" asks that this treaty be refined. Also, unlike interstate compacts, treaties do not have a firm judiciary or higher authority backing them up. Diplomacy is required, implying the presence of institutional uncertainty, as diplomacy is slow and has unpredictable outcomes.

An interesting observation of consequence to other jurisdictions concerns the manners in which these compacts and treaties were obtained. What brought governmental representatives to the table and got them to design and sign successful agreements? First, some of these agreements were the results of many years of labor. So they are not easy tools to forge. Second, funds from outside Texas played a major role in enabling these useful instruments. Some compacts were directly stimulated by states' desires to benefit from federal water projects that could not be initiated until states resolved differences over water sharing (Chapter 8). Even the 1944 Treaty has a strong indebtedness to U.S. commitment to develop large reservoirs on the Rio Grande, two of which were subsequently built. Hence, in situations where jurisdictional divisions of water are sorely needed, look for opportunities to buy action by dangling carrots. The benefits of these agreements can last for a long time and better justify the expense.

The Trials of Environmental Stewardship

Decisionmaking in Texas has demonstrated a strong sensitivity to prospective profit gains and losses, perhaps more so than in most other U.S. places. Creating wealth, enhancing property value, increasing the tax base, and generating jobs are almost always viewed as positive directions in the hearts of Texas policymakers, so it has been easier to make public policy choices that are supportive of these things. Policies that might hamper these most evident expressions of growth are poorly received. Development interests have been quite willing to champion their proposals in this decisionmaking environment, because their lobbying efforts can be successful.

On the other hand, the diverse burdens of development—the social cost side, as might detract from the quality of life or the environment—have not been similarly represented or heard during much of Texas's water history. The possibility that growth no longer aids progress has not caught on, as lay opinion maintains the belief that further economic development promotes public welfare.[1] Part of the reason may lie in the fact that the costs of growth fall broadly across the Texas population. With small costs per individual, citizens are less motivated to take action politically or even to seek a better understanding of what might be going on. Part of the reason for underrepresentation of costs certainly lies in the scientific difficulty of knowing the linkages between such things as water use and environmental consequences (Chapter 6). Even where the existence of a connection becomes routinely accepted—an accomplishment in and of itself— lack of precision over the exact nature of a linkage provides a measure of deniability favoring still more development. This "inertia of understanding" appears to condemn us to a path whereby we overexploit our water resources, to the point of causing too much damage to ecosystems and harm to established water users before corrective attention can be marshaled.

Because of these forces, it is easy to understand why an extraction-oriented favoritism arose in Texas. Among the results is that threshold instream flows remain unestablished. Consequently, too many or too senior surface-water rights have been issued to private parties in some basins, and administrative processes overseeing additional water development, water exchanges, and interbasin transfers have been overlooking an important detail. It appears that a more accurate balancing of relative social values would leave more water in particular streams and rivers. Similarly, groundwater pumping rules seem to have been too permissive in light of the springflow losses that have occurred.

What can be done at this late stage? In some ways, "the horse is out of the barn," as a Texan might say. Are we to somehow revoke a portion of the water rights that were granted to so many surface-water and groundwater users? Given the political tendencies already noted, why would policymakers do this?

The short answers to these questions, as demonstrated in Texas, are that it is tough to change course, but new administrative procedures can be adopted, at least for administering the approval of new water rights and future water reallocations (Chapter 7). Additionally, water marketing provides a sanctioned, though possibly expensive, method for retrieving water rights. Texas recently commenced doing some of these things. Why the beginnings of change? It must be admitted that a lot of this change has been induced by nongovernmental parties. Texas courts are obligated to respect precedents and tend to be locked to the aged institutions that are failing to keep pace with water scarcity. The Texas legislature remains very sensitive to private development opportunities, and it is not enthusiastic about rededicating water to the environment, nor is it ready to question the merits of still further growth.

The revolutionary reform adopted by the legislature for the Edwards Aquifer was initiated by a federal judge hearing a Sierra Club suit against the state (Chapter 5). On being forced to establish a new regime for the Edwards, it was natural for the state to turn to a transferable, quantified rights alternative, a system with which it is familiar for surface water and land. The new approach appears to be providing good overall service.

In the case of policy change recognizing instream flow and estuarial inflow needs, small policy seeds planted in the 1980s were slowly growing, but momentum did not actually emerge until the 2000s, when environmental groups attempted to claim large new water rights in multiple basins (Chapter 7). The resulting stimulus for policy change is quite evident, although it is premature to say whether the policy outcomes will be a significant new course. And the tricky issue still remains of bettering conjunctive groundwater–surface water management and its environmental deficiencies. Until groundwater management can mature to the point of respecting surface water influences, especially springflows, environmental failures will persist.

Planning's Admiration of Structural Approaches

Texas has more than 40 years of experience in authoring state water plans every several years (Chapter 2). The resulting plans are consistent in content and delivery.

Each published plan inventories current use information by sector, locale, and water source. These plans project growth in water use (based primarily on extrapolation of population trends), find that projected use exceeds ability to supply,[2] and call for generally massive infrastructural investments. Beginning with the 1984 edition, these plans began to acknowledge the desirability of soft approaches toward water scarcity, referring to them collectively as "conservation," but the plans nevertheless have continued to place infrastructure "needs" front and center.

In the late 1990s, Texas inverted its planning approach in a highly trumpeted move to "bottom-up" planning. The state was divided into 16 planning areas, and citizen planning groups were formed (Chapter 2). The groups hired consultants and received informational support from the state's water-planning agency. Work from the 16 groups was combined to form the next water plan, published in 2002, and the process was repeated for the 2007 plan.

Judging from the emphasized results of the 2007 plan, Texas still believes that water supply development is key to aligning the state's growth with its natural water supply.[3] Overall, these plans say that if we could just transport water to the right places, or capture it so that it does not get wasted upon the sea, or make use of untapped water resources such as brackish groundwater, Texas may be able to withstand the growing problems.[4] Interestingly, the newer planning approach highlights the same solutions as did the top-down approach. The planning demeanor remains focused on making more intensive use of available water, as opposed to living within the means of currently developed water. The ideas that further development might sacrifice valuable environmental resiliency or that water use should be retuned to match supply, rather than the other way around, have yet to gain much support.

Technological Options

Another supply-side path to consider is the possibility of technological rescue. Several noteworthy technology strategies are called upon in regional planning documents. Many are in practice in the state, and the possibility of enlarging these activities is being promoted. The opportunities here are wide-ranging, including various types of desalination, water reuse, and water harvesting. These are inventoried in Chapter 10; for each alternative, features are described and the level of utilization in the state is identified. Major desalination plants are now operating in the state, and some cities are engaged in water reuse. Rainwater collection has attracted considerable public interest.

Desalination is the elephant among these opportunities, in that it alone has the ability to bring substantial new supplies to bear on the scarcity problem. Reuse and water harvesting extend local supplies, but with a tendency to reduce downstream water supply, implying that they are not strictly supply-enhancing measures. By bringing brackish groundwater or Gulf of Mexico seawater into usability, the various desalination technologies have the potential to increase potable water supplies without lessening downstream supply. However, there are still challenges to consider (Chapter 10). Capital costs are high, energy use can be large (implying sensitivity to energy prices), and the environmental harms of salt disposal can be prohibitive or can raise costs. Present approval procedures do not require that desalination projects pass economic tests, so it remains unclear whether they are making net social contributions. In light of these hurdles, the experience being generated by Texas desalination facilities promises to illuminate these options further in the years ahead.

Marketing's Breadth and Reach

The force of water marketing as an instrument of change has been strong in Texas. At least five distinguishable markets exist (Chapter 4). They have been responsible for the exchange of very large amounts of water and have enabled some regions to handle population growth inexpensively, relative to alternative water supply development costs. The benefits of these accomplishments to the people of Texas have likely been quite high.

New water markets can be created to deal with long-standing problems. When confronted with the realities of forced habitat protection for Edwards Aquifer outflows, the legislature created new water rights and thus a new marketplace, which quickly became vibrant and high-volume.

Groundwater access marketing is a form of water marketing. Although it has helped in some ways, it has major shortcomings. When groundwater can be exchanged only through land-based transactions—as everywhere in Texas except for the Edwards Aquifer—groundwater marketing is not reaching its potential. Until actual water rights are quantified and enforced, groundwater will remain difficult to allocate well. Yet this is also a problematic transition, ordinarily requiring assignment both of water that is in storage and of water that regularly enters aquifer systems as recharge.[5]

One of the five markets has enabled the monopolistic growth of river authorities in eastern Texas basins. Perhaps river authorities can accomplish good service (and reallocation to higher-valued uses) with their large-scale acquisitions of water. Perhaps they can improve instream flow management within their areas too. Perhaps they can manage their internal conduct so that the net benefits derivable from water, and, it is hoped, received by water users, do not get expended within the river authorities themselves, in the form of either unnecessary expenditures or inefficient decisions. Yet the power of river authorities stymies water marketing among lesser water users. In basins where river authorities are strong, small-scale transactions are rare among ordinary water users, and planners within these basins are deprived of a crisp observation about the value of water and how this value depends on conditions such as location, season, or weather.

Although progress is slow, it is comforting to witness the role of water markets in helping the value of natural water get reflected in retail water rates. The absence of water value in water rates is not just a Texas problem, so maybe policy models exist elsewhere that can assist with this issue. When communities acknowledge the intrinsic worth of the water rights they control and cause this value to be present in water rates, people are motivated to behave more appropriately as they make water conservation and use decisions.

Interbasin Transfers

One of the most contentious provisions of current surface-water law is the relatively new junior water rights provision (Chapter 3). Under this 1997 rule, any interbasin transfer of surface water becomes junior to all existing water rights in the originating basin. Ongoing debate suggests that this rule may be dropped or moderated by future legislation.

On the one hand, this rule lowers the value of interbasin transfers, thereby discouraging a specific class of water marketing, even in situations where potential buyers value water more highly than sellers. On the other hand, this rule maintains in-basin flow regimes that are more intact; reduces pipeline costs, land condemnation, and environmental disturbances; and asks that each basin live within its natural water endowment. Supporters argue that it preserves rural opportunities for economic development.

Whether the gains of the junior water rights provision exceed the losses is an interesting, unanswered question. However, such a question presumes that legislative intent might involve equal weighting of originating and receiving basins. Perhaps the 1997 (and later) legislatures have been expressing their beliefs that additional weight is deserved for basins of origin. If this is the case, it is an understandable expression of social goals.

Local Groundwater Administration

In an effort to manage groundwater in a more modern way without imposing the state's will, the legislature has encouraged the formation of locally controlled groundwater districts, of which nearly 100 now exist (Chapter 3). As they are prone to be defined using political boundaries rather than hydrologic ones, there are numerous one-county districts.[6] They are allowed broad discretion in the policy measures they employ, so they may produce interesting data concerning policy choices, justifications, and results. Their policy developments are being watched by many observers in the state.

Whether districts can successfully patch the leaky rule of capture is an unanswered question (Chapter 4). Whether they can improve rates of depletion, the efficiency of water use, and the interface between surface-water and groundwater law remains open too. Their attempts are not made without substantial difficulty. If the districts are not aggressive, not much can improve, as shown by the Ogallala analysis in Chapter 9. But when they try to think differently and get control of water use, they risk expensive conflict. When a Hudspeth County groundwater district recently attempted to quantify rights, for the purposes of aligning use with available supply and enabling groundwater exports to an urban area, its efforts were found by the Texas Supreme Court to mistreat landowners who had little history of water use (and arguably a long history of conservation).[7] Where this 2008 ruling leaves districts is unclear. Can quantification of groundwater rights be conducted in a way that is fair and allows for export? If quantification of groundwater rights is not feasible, how can these institutions evolve and how can groundwater management ever become coordinated with surface water policy?

Arguably, the legislative choice to pass responsibility to groundwater districts was the easy direction to take. It would seem that local authorities should be able to deal with a local problem. Yet this path also appears too incremental and timid, at a time when a more forceful departure from the rule of capture was long overdue. A fundamental matter is that the consequences of the pumping problem are not entirely local. Some aquifers are large; none respect political boundaries. Most aquifers are hydrologically connected with surface water, implying basinwide as well as environmental consequences. The welfare of future generations is at stake because of groundwater depletion. These nonlocal dimensions call into question whether a local political authority has either the motivation or the capacity to render socially advantageous policy. Moreover, in contrast with scores of independent districts, uniform state policy would be more likely to have lower management costs through standardization of rules, oversight, and data collection.

AN INTERESTING FUTURE

The several frontline issues being encountered in Texas are not likely to be resolved soon. To a large extent, they are ingrained now. By their nature, they constitute considerable challenges as a consequence of embedded trade-offs lying at the core of water resource decisionmaking: agrarian transformations to urban, upstream versus downstream, groundwater users versus interconnected surface-water users, present people versus future people, environment versus diversions.

The free lunches of the original Texas water endowment have been consumed. Now every policy intended to direct additional water to a group of specific water users will necessarily reduce the water available to others. Surplus water exists in wet-weather cycles, when people are less apt to want it, but normal-period surpluses can be found only in far eastern parts of the state. The large number of people residing in Texas and the advanced state of economic development have led to water overemployment. At this intense level of water use, hydrology keeps all water users entwined. Even swaps of water among users can have implications for third parties because of hydrologic intricacies and flows. As a consequence, changes in one user's water consumption really do affect others, so every imaginable policy modification promises that some people will be aided while others are harmed. New dams and pipelines do not manufacture water; they redirect it. The easy adjustments have been made; only the hard ones remain.

NOTES

1. The history of this ideology as well as its reality and the connectivity between growth and happiness are intriguing to consider. See Peter A. Victor, *Managing without Growth: Slower by Design, Not Disaster* (Northhampton, MA: Edward Elgar, 2008).

2. Such findings are unavoidable because these demand projections are performed without accounting for rising scarcity. According to economic principles, excess demand (when demand is greater than supply) is a necessary outcome when the rising value of a resource is omitted from planning methodology.

3. Texas Water Development Board (TWDB), 2007 State Water Plan, Chapter 1, http://www.twdb.state.tx.us/publications/reports/State_Water_Plan/2007/2007StateWater Plan/CHAPTER%201%20FINAL%20113006.pdf (accessed June 2, 2010).

4. It is educational to observe that the 1968 plan called strongly for water importation from the Mississippi Valley. Grand project proposals such as this one have not gotten very far.

5. V.L. Smith, Water Deeds: A Proposed Solution to the Water Valuation Problem. *Arizona Review* 26 (1977): 7–10. Yet the flow-dominated nature of the Edwards Aquifer implies that transfers of flows (annual recharge) achieve a great deal without requiring ownership assignment of water already in storage.

6. TWDB, Groundwater Conservation Districts, 2009, http://www.twdb.state.tx.us/mapping/maps/pdf/gcd_only_8x11.pdf (accessed June 2, 2010).

7. Supreme Court of Texas, *Guitar Holding Company v. Hudspeth County Underground Water Conservation District No. 1, et al.*, 2008, http://www.supreme.courts.state.tx.us/historical/2008/may/060904.pdf (accessed June 2, 2010).

Index